# Hot Rod Body
# and Chassis Builder's Guide

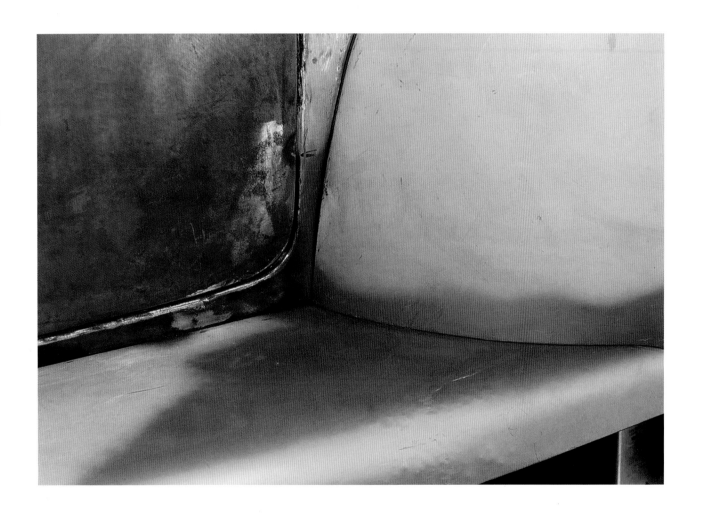

*By Dennis W. Parks*

# Dedication

To Joan and Ed Thornton, my aunt and uncle, who took me to my first car show so many years ago.

First published in 2009 by Motorbooks, an imprint of Quarto Publishing Group USA Inc., 400 First Avenue North, Suite 400, Minneapolis, MN 55401 USA

Motorbooks titles are also available at discounts in bulk quantity for industrial or sales-promotional use. For details write to Special Sales Manager at Quarto Publishing Group USA Inc., 400 First Avenue North, Suite 400, Minneapolis, MN 55401 USA.

To find out more about our books, join us online at www.motorbooks.com.

ISBN-13: 978-0-7603-3532-1

Editor: James Manning Michels
Design Manager: Jon Simpson
Designed by: Pauline Molinari

Printed in China

**On the cover:**
This classic hot rod was built from a 1930 Ford Model A five-window coupe by Richard Graves. *Peter Harholdt*

**On the title page:**
Weld-through primer keeps the sheet metal clean but can be welded through without first grinding or sanding it off.

**On the back cover:**
It looks like this body and frame have been stored in a dry climate for quite some time. Having the body chemically dipped or media blasted and then coating it with epoxy primer would make this vintage tin look great in no time.

This pneumatic angle grinder is set up with a sanding disc, but could also be equipped with an abrasive disc for cutting and grinding. Using the right tool makes any job easier.

**About the author:**
Dennis Parks' many books include *How to Build a Hot Rod, How to Build a Cheap Hot Rod, How to Build a Hot Rod Model A Ford, How to Restore and Customize Auto Upholstery and Interiors,* and *How to Paint Your Car,* all from Motorbooks. Parks lives near St. Louis, Missouri. dennisparks.com

# Contents

# Foreword

So, you want to build a hot rod . . . You might ask, "How do I start this process?" Well, the answer is in your hands right now. This book will provide critical information for planning and executing the project. The book starts with explanations of the tools and equipment needed. It will give you information on the different styles and types of suspensions and valuable technical information about how to set these up correctly. It then continues on through the build process, with many detailed tips and explanations. As you work on your project, you can use this book as a resource for technical information and to look at various options and alternatives as you proceed.

The beginning of any project is important. If a carpenter starts a house out of level or out of square, nothing he does after that will ever be correct. A lot of work will be done to try to correct the original error, and the house will never be as it should. This is also true for hot rods. Make every effort to prepare yourself before you begin and to make each step right before you continue.

When I worked as an industrial arts teacher, my advice to students was that "proper prior planning prevents poor performance." Think ahead, look at the options you have, and go with the best choice that meets your objectives. Gather together as much information as possible on the individual parts of the build. Have an overall plan for the project and additional plans for each individual phase. You won't build the car all at once, but in steps. Your overall plan keeps your end goal in focus, and your individual plans keep the project moving forward, one accomplishment at a time.

First, look at what your objectives are for the car you are building. How it should look, the ride height, the ride quality, and the handling are just some items to consider. Much of the time it is a tradeoff, as you have to sacrifice something to get something else. The important part is to think and plan ahead of time. Choosing an overall look and feel, and components that work with that ideal, will save you a lot of time and money.

Often, as a build goes on, you have to make adjustments to your original plan. If you start with a good plan, it will make it easier to make these changes, and you may have already thought of the other options you have. The frame and the suspension begin to set the theme of the car you are building, and you want to continue that theme throughout the build. So, use this book to plan the build of your hot rod. Think ahead, and if you do need to change or modify your plan, you can use the information in this book to steer you in the right direction. The technical information contained here can be an invaluable bonus.

**John Kimbrough**

# Acknowledgments

As much as this book deals with hot rod chassis and bodies in particular, it deals with hot rods in general just as much. During my writing sessions, I found myself questioning why certain things that I knew to be true were that way. I had to check my sources as to why and rethink some conventional wisdom so I could explain it in a way that makes sense. Hopefully, this extra work makes the explanations here as simple and as logical as they can be. Building hot rods is not necessarily a complicated proposition, but like most talents, you'll quickly lose your edge if you don't continue to practice your craft.

In addition to a lifetime of wrenching on and building cars, I have the good fortune of knowing a vast group of hot rodders who have built, are building, and will continue to build hot rods. Luckily for me, they will usually answer my questions, no matter how many times I ask. In no particular order, a great big "thank you" goes out to Ed Thornton, Bob Galbraith at New Port Engineering, Keith Moritz at Morfab Customs, Don Cain at KC Street Rod Parts, Roger Ward at Bad Paint Company, Jerry and Jason Slover at Pete & Jakes Hot Rod Parts, Danny Miller at Rear Gears, Kevin and Wendy Brinkley at the Paint Store, John and Sam Kimbrough, Tom Prufer, Jack and Donnie Karg at Karg's Hot Rod Shop, Vince Baker at KBS Fabrications, Tim Kohl, and Steve Gilmore. All of these guys contributed to this book in one way or another. For some, it was during time spent walking around countless rod runs, others while I was photographing work being done in their shop, and still others during numerous phone calls, e-mails, and shop visits. Again, thank you for sharing your collective wealth of information. May we all continue to build hot rods for many years to come and feel some satisfaction that we have all done our part to pass the torch.

**Dennis W. Parks**

# Introduction

**M**y previous book, *The Complete Guide to Auto Body Repair,* dealt with the sometimes necessary skills required to repair body damage. Most of the skills and techniques that were discussed in that book address standard repair issues that are typically encountered with your ordinary daily driver. There is some creativity in how you repair something, and certainly taking a beautiful car that has been crunched and restoring its original lines is satisfying. . . . Yet repair work isn't really the stuff of dreams. Hot rod building is!

*The Hot Rod Body and Chassis Builder's Guide* should stir some excitement and creativity in your mind. Once you learn the various methods of fabrication and how to work with more advanced tools, your hot rod projects will be limited only by your imagination.

In this book, Chapter 1 discusses the various tools that are typically found in any competent hot rod fabrication shop. Bear in mind that you do not have to own all of these tools to build a cool hot rod, but having the correct tools for the job and knowing how to use them properly will always make the task at hand easier. Chapter 2 concentrates on chassis work, whether you are resurrecting an original chassis for use beneath a high-horsepower hot rod or starting from scratch with a pair of bare frame rails. Taking a production

body down to a decent starting point is the goal of Chapter 3, where body repair is discussed. No matter how pristine that new project may be, the body most likely needs some repair, be it rust removal, a small patch panel, new floors or firewall, or just a few (or several) holes filled. Chapter 4 discusses reproduction bodies and some of the things that you need to know, whether the body is fiberglass or steel. By the time you spend the coin required to bring home a reproduction body, you should make sure that you know all there is to know about making that body last. Chapter 5 is where the fabrication really begins and things get quite exciting. Not that the rest of the book is anything to sneeze at, but this chapter should get your creative juices flowing. Chapter 6 discusses the various tasks that are required prior to applying the perfect finish to your newly created work of art. Chapter 7 provides some visual ideas and discussion of various paint schemes, along with how to implement them on your hot rod. These last two chapters also show yours truly actually doing some work on my track T, practicing what I'm preaching here. If I can do it, you can!

I hope you enjoy this book, and thank you for buying a copy.

Whether it is an artist's sketch, a computer-enhanced drawing, or a picture from long ago etched forever in your mind, having an image of how you want your hot rod to look in its finished form will be helpful in meeting that goal. When you have nothing to guide your work, you will find yourself varying your ideas with each new fad or suggestion from someone else. *Artwork courtesy of Stilmore Designs*

# Chapter 1
# Tools

**T**ools . . . just where would we be without them? No doubt, we'd probably have more coin in our pocket if we were not paying for the massive roll-around tool chests that are chock full of the latest "must haves" from Snap-on, Mac, or Craftsman. But, if you are reading this, you must be a working man. You probably cannot imagine not working on a hot rod at least part of your working day and would prefer to be welding a chassis together at two in the morning than working in the plushest office in town . . . ever.

No matter what your occupation may be, there are certain tools required for that line of work. Those required for hot rod chassis and bodywork just happen to be way cooler than anything an accountant or lawyer may use. Some of these tools are pretty basic and therefore mandatory, while others are more luxurious. Before plunking down your hard-earned money for the latest offering from the local tool emporium, you should consider a few basic questions. Do you already have an acceptable method of doing what the new tool or piece of equipment will do for you? How often do you perform that task? Will the new piece of equipment save enough labor time to justify the cost? Most tasks that can be performed with power tools can be done by hand, although power tools will usually save time. If you are running a production shop, that time saved translates into money made; if you are working by yourself in your backyard garage, it may be less important.

Having the right tools for the job and knowing how to use them correctly is the key to success in any profession. Even if you have to improvise, simply knowing which tool will serve as a better alternate will yield a better product. Whatever tool you are using, please be sure to familiarize yourself with its operational and safety aspects. Building hot rods should be fun, and I don't recall a trip to the emergency room ever being fun.

## PNEUMATIC TOOLS

Not all tools involved with metal fabrication are pneumatic, but a large number of them certainly are. From cutting to smoothing, and on to priming and painting, compressed air is the typical power source for most of the tools designed for doing these chores.

Pneumatic tools do not require a motor, as the air compressor converts electric energy to kinetic energy. This allows pneumatic tools to be smaller, lighter, and therefore less cumbersome to handle. For this same reason, they take up less space when they are not being used. Additionally, they typically provide more torque and can run at higher rpms, allowing them to perform their chosen task more quickly and efficiently.

Besides the initial expense of an air compressor, air-powered tools are typically less expensive than their electric-powered counterparts. So once you have a compressor, having a wider variety of pneumatic tools is more feasible while staying within a prescribed budget. Most air tools require a certain amount of lubrication, so you should consult and follow the maintenance information provided with each air tool for best service.

Air-powered tools typically run quieter and cooler than those powered by an electric motor. To some people, this might not seem like a big deal, but as the number of tools in use at one time increases, you can certainly notice the difference. More quiet and coolness makes for less fatigue.

### Air Compressor

When shopping for an air compressor, you should consider two basic components: the pressure in pounds per square inch (psi) and the volume in cubic feet per minute (cfm). A motor is what actually compresses the air, so there will be a horsepower rating given—but by itself, the rating is really meaningless. Typically, the greater the horsepower, the greater the pressure that can be achieved. What is important to remember is how these two ratings perform together. An air compressor that can deliver 10 psi at 1 cfm can inflate an air mattress, but will take quite a while to do it.

As a standard, most air compressors are rated at 40 psi or 90 psi, and sometimes both, such as 4 cfm at 90 psi or 12 cfm at 40 psi. This provides the constant when you are shopping for a compressor. The variable is how much air volume you need. A typical impact wrench may require 4.5 cfm at 90 psi, while an HVLP (high volume, low pressure) spray gun may require 8.5 cfm at 40 psi. The first air compressor (4.5 cfm at 90 psi) would not be able to provide enough volume to run either of these tools efficiently, while the second air compressor (12 cfm at 40 psi) would be able to power either.

All things being equal, the more psi that a compressor can generate, the more volume it can deliver. However, if the size of the air storage tank is too small, this will restrict the efficiency of the compressor. In other words, if you are using a tool that requires lots of volume, such as an HVLP spray gun, the volume of the tank may become a more critical factor than the air pressure rating.

An air compressor motor operates to fill the air storage tank, and when the tank becomes full, the compressor shuts off. As compressed air is used or bled off, the compressor begins to run again, repeating the cycle. The more the air compressor motor runs, the hotter it makes the air that is being compressed into the air storage tank. As this hot air cools, it condenses into a liquid (water). If the tank is large enough to store enough air to complete the task at the appropriate air pressure, the air compressor motor will not be required to run, enabling you to use cool, dry air. If the tank is too small to store enough volume, the compressor motor will be forced to run more often, which will cause the compressed air to be warmer and more humid. Due to this creation of moisture, the air storage tank should be drained on a regular basis. This will help prevent rust in the tank, moisture damage to your air tools, or water spitting out and ruining your paint job.

### Regulators

Air pressure is adjusted by tightening or loosening the knob on the air compressor's regulator. Smaller, homeowner-type compressors typically have a built-in regulator, while larger compressors usually have a separate regulator plumbed into the air line that exits the compressor's air storage tank. Some shops may have multiple regulators in the air line. In a general purpose shop, one regulator adjusted to provide sufficient pressure for the particular tool that requires the most air would be adequate. In a larger shop that includes mechanic work, priming, and painting, each area would need a regulator to control the air pressure.

If you are going to be using pneumatic tools, you are going to need plenty of compressed air. This commercial unit requires 240-volt electric service and has about a 60-gallon air tank. For use in your home garage, you may be better off with a 110-volt unit, unless you have extra electric service installed. Whichever you choose, get the largest tank you can afford and have room for.

Pressure regulators are essential between the air compressor and the tool being used. This particular model has an air supply line entering from the right and is used to regulate two separate lines that exit from the left. Sanders, grinders, and air saws require a certain amount of air to operate properly, while spray guns must be operated within a specific air pressure range, depending on the material being applied.

In addition to regulators that control the flow of air to the entire air line downstream, there are also regulators built into many spray guns. These work in much the same way, but provide the user more accurate control when spraying.

Regulators are often designed to include driers and filters as an all-in-one air management unit. These can be pricey, however, so if you are just starting to equip your shop, you may choose to add driers and filters separately.

### Driers

Remember that condensation that I mentioned a few paragraphs ago? You have to get rid of it somewhere. Of course, the best way is to drain the air storage tank on a regular basis. However, even if you have drained the tank at the beginning of the workday, a full day of compressor use will create some amount of moisture, depending on ambient temperature and humidity. Since moisture is an enemy of air

Shown is a Sharpe filter/drier/regulator. A copper hard line supplies air from the compressor. The compressed air then passes through the filter/drier to eliminate moisture or other contaminants, and then can exit from either of two regulated outlets.

tools and can absolutely ruin a paint job, installing a drier in the air line somewhere between the air compressor and the outlet will increase the life of your air tools and potentially save hours of painting effort.

Most driers contain a desiccant material that soaks up moisture. This eliminates the need to drain moisture from the drier; however, the desiccant will sometimes need to be replaced or reactivated.

### Filters

In addition to eliminating moisture, it is good to filter out anything else that might be floating around in your air supply. This includes the rust that forms on the bottom of the tank, since you forgot to drain the tank on a regular basis. Filter housings are installed inline much like regulators and driers. In addition to the air line connections, the filter housing includes a cup that can be removed. The cup is where the actual filter is located. The filter can be removed, the cup wiped clean, and the filter cleaned or replaced.

### Hoses

Air hoses are the flexible link between your air compressor and the air tools that you are using. Besides having the proper fittings on both ends so that there is no air leakage, air hoses must be of the proper size to work efficiently. Many homeowners use ¼-inch self-coiling hose, as it keeps itself

**Minimum Pipe Size Recommendations**

| Compressor Size | Capacity | Main Air Line Length | Size |
|---|---|---|---|
| Up to 2 hp | 6 to 9 cfm | Over 50 feet | ¾ inch |
| 3 to 5 hp | 12 to 20 cfm | Up to 200 feet | ¾ inch |
| | | Over 200 feet | 1 inch |
| 5 to 10 hp | 20 to 40 cfm | Up to 100 feet | ¾ inch |
| | | Over 100 feet to 200 feet | 1 inch |
| | | Over 200 feet | 1 ¼ inch |
| 10 to 15 hp | 40 to 60 cfm | Up to 100 feet | 1 inch |
| | | Over 100 feet to 200 feet | 1 ¼ inch |
| | | Over 200 feet | 1 ¼ inch |

The size and capacity of your air compressor along with the distance the compressed air must travel from the compressor to the air tools that it operates is a determining factor on what size pipe should be used. Copper or galvanized pipe should be used to plumb your air supply to where it will actually be used.
*PPG Finishes*

coiled up nicely when not in use. However, that kind of hose simply isn't going to pass enough air to operate the types of pneumatic tools that will be used in a chassis shop or body shop. Although you may be able to skimp by with a 5/16-inch inside-diameter hose, one with a 3/8-inch diameter would be a better choice. Simply put, even if your air compressor can provide enough pressure and volume, using an air hose that is too small will hamper your capabilities. You should also limit the length of your workspace's air hoses to 25 feet; otherwise you will experience significant pressure loss due to friction through the hose.

To minimize the required air hose length, consider installing an air supply system made of 3/4- or 1-inch (or larger for commercial applications) copper pipe or galvanized pipe. This can be surface mounted to the inside walls and ceiling of the shop, so it is never too late to install it, regardless the age of the building. This size piping will not experience the friction loss that is common with smaller diameter hose. Multiple quick disconnect plugs can be plumbed into the system at convenient locations so that air hoses are not running all the way across the shop floor.

## Air Pressure Drop at Spray Gun

| Size of Air Hose (Inside Diameter) | 5 ft length | 10 ft length | 15 ft length | 20 ft length | 25 ft length | 50 ft length |
|---|---|---|---|---|---|---|
| **1/4 inch** | Air Pressure Loss in psi | | | | | |
| at 40 lbs. pressure | 06.00 | 08.00 | 09.50 | 11.00 | 12.75 | 24.00 |
| at 50 lbs. pressure | 07.50 | 10.00 | 12.00 | 14.00 | 16.00 | 28.00 |
| at 60 lbs. pressure | 09.00 | 12.50 | 14.50 | 16.75 | 19.00 | 31.00 |
| at 70 lbs. pressure | 10.75 | 14.50 | 17.00 | 19.50 | 22.50 | 34.00 |
| at 80 lbs. pressure | 12.25 | 16.50 | 19.50 | 22.50 | 25.50 | 37.00 |
| at 90 lbs. pressure | 14.00 | 18.75 | 22.00 | 25.25 | 29.00 | 39.50 |
| **5/16 inch** | | | | | | |
| at 40 lbs. pressure | 02.25 | 02.75 | 03.25 | 03.50 | 04.00 | 08.50 |
| at 50 lbs. pressure | 03.00 | 03.50 | 04.00 | 04.50 | 05.00 | 10.00 |
| at 60 lbs. pressure | 03.75 | 04.50 | 05.00 | 05.50 | 06.00 | 11.50 |
| at 70 lbs. pressure | 04.50 | 05.25 | 06.00 | 06.75 | 07.25 | 13.00 |
| at 80 lbs. pressure | 05.50 | 06.25 | 07.00 | 08.00 | 08.75 | 14.50 |
| at 90 lbs. pressure | 06.50 | 07.50 | 08.50 | 09.50 | 10.50 | 16.00 |
| **3/8 inch** | | | | | | |
| at 40 lbs. pressure | 01.00 | 01.25 | 01.50 | 01.75 | 02.00 | 04.00 |
| at 50 lbs. pressure | 01.50 | 01.75 | 02.00 | 02.75 | 03.50 | 04.50 |
| at 60 lbs. pressure | 02.00 | 02.50 | 02.75 | 03.50 | 04.25 | 05.00 |
| at 70 lbs. pressure | 02.50 | 03.00 | 03.25 | 04.00 | 04.75 | 05.50 |
| at 80 lbs. pressure | 03.00 | 03.50 | 04.00 | 04.50 | 05.25 | 06.00 |
| at 90 lbs. pressure | 03.75 | 04.25 | 04.75 | 05.25 | 05.75 | 06.50 |

The inside diameter of air hoses can affect the amount of air pressure delivered to a paint spray gun. This chart shows some basic pressure drops for 1/4-, 5/16-, and 3/8-inch inside diameter air hoses when used at specific lengths. Keep these calculations in mind when determining the correct pressure for spraying undercoats and paint so spray application will be made at recommended gun pressures. *PPG Finishes*

## Cutting Tools

Metal cutting tools have improved considerably in the last 60 years or so. Where a cutting torch was once the norm, contemporary hot rod shops are equipped with plasma arc cutters, pneumatically operated shears, nibblers, air saws, die grinders, and air chisels. Each of these has its own primary and secondary uses, in which they are the best tools for the job. Perhaps the biggest advantage that today's fabricators have over their predecessors is that contemporary cutting tools typically offer a more precise and smoother cut than those of yesteryear. Anyone who has ground smooth the edges of anything cut out with a cutting torch can recognize the high quality and convenience of a plasma arc cutter.

### Plasma Arc Cutters

Priced around $1,500, a plasma arc cutter may be the best bang for your buck when it comes to equipping your shop. Some models will cut through thicker materials than others, so you should have a good idea of what maximum thickness of material you will be cutting before you make your purchase. Whether you are cutting sheet metal or heavier stock (bars, channels, and angles), a properly sized plasma arc cutter makes the job an easy task. A drawback with plasma arc cutters is that they are not efficient for cutting through multiple layers of material.

Plasma arc cutters require compressed air and electricity to operate. Several models are available that utilize

This plasma arc cutter from Thermal Dynamics can be wired to run on 110 or 220 voltage and has an electric cord extending from the back. Also in the back of the unit is an air nozzle to which an air hose can be connected. This particular unit requires about 60 psi of dry air along with the aforementioned electric supply to operate. The large alligator clip attaches to the work surface to serve as a ground.

110/115-volt electric service. These are typically capable of cutting material up to about ⅜-inch thick, which is about as thick as any material found in a hot rod shop. Larger, more expensive models that require 230-volt electric service are capable of cutting material up to 1 ¼-inch thick. You most likely won't need to cut anything that thick, but if you need to, there's your answer.

Besides the rated thickness, or rated capacity, that a plasma arc cutter can cut, you may hear of its "maximum pierce capacity." The rated capacity is based on starting from an edge and cutting across the material, while the pierce capacity is the ability to bore through material when not starting at an edge (similar to drilling a hole in the middle of the material). The maximum pierce capacity is typically one-half of the rated capacity.

A plasma cutter utilizes a cutting head that resembles a MIG welding torch. You simply place the cutting head on the material to be cut, squeeze a trigger or push a button, and then pull the cutting head along the line to be cut. While a more precise cut can be made with a template or guide, plasma arc cutters can be used freehand. Of course, the best way to proceed depends on how steady-handed you are.

When using a plasma arc cutter or welder around glass (or flammable materials), you should use a welding curtain so that any sparks given off don't cause glass damage or a fire. Plasma arc cutters use an incredibly hot arc to melt the material they are cutting, so keep your fingers clear. Wearing proper and properly fitting gloves is a good idea with high-temperature tools. Note that there are some spinning tools you can find in a shop where gloves may not be a good idea because the tool can get hold of it and torque fingers very hard, very fast. Read the tool manufacturer's operating and safety instructions, and proceed accordingly.

### Shears

Pneumatic metal cutting shears operate much like a pair of air-powered scissors. The air supply hose connects near the base of the handle, which has a pistol grip. At the opposite end are two (or sometimes three) blades that work against each other to cut sheet metal without distorting it. Pneumatic shears can be used to cut freeform shapes in metal up to approximately 18 gauge. Different types of shears are discussed later in this chapter.

### Air Saws

An air saw is nothing more than a reciprocating saw blade that is powered by air. If you are familiar with a scroll saw, a jigsaw, or reciprocating saw, you can quickly recognize this as being able to cut freeform shapes. Air saws can be equipped with blades of different tooth counts to provide smoother cuts (more teeth) or faster cuts (fewer, bigger teeth). A benefit of an air saw is that it can cut in confined spaces, while a drawback is that blades can be bent easily. Air saws are great for trimming and cutting fiberglass.

Being portable and accepting cutting blades with different tooth counts, an air saw becomes a very versatile tool when you are working with sheet metal. Very intricate shapes can be cut with the relatively narrow blade.

An angle-head grinder works the same as an ordinary die grinder, but it is better able to reach into tight spaces. This one is set up with a sanding disc, but could easily be equipped with an abrasive disc for cutting and grinding.

### Die Grinders/Cut-Off Tools

Although they operate the same way and perform the same task, die grinders are available in two distinctly different configurations. In both types, the air hose attaches to the end of the die grinder's body, the body serves as the handle when in use, and a lever-type trigger is squeezed against the body to operate it. The difference in the two styles is that on one, the grinding wheel rotates about an axis that is parallel to the body of the die grinder, while on the other the grinding head is mounted at 90 degrees to the body. If you have plenty of room in which to work, this is not a big deal, but when space is limited, a die grinder with the angled head will usually be more maneuverable.

For making relatively straight cuts in sheet metal, such as when cutting out a rusty area prior to installing a patch panel, a die grinder with a cut-off wheel works very well. A die grinder can also be used as a cut-off tool for cutting tube or bar stock to length.

With different accessories, such as a sanding pad, an angle-head die grinder can be used to sand or grind in confined areas, rather than cutting. Additional accessories that can be used with these tools are deburring tips, grinding stones (in a variety of shapes and sizes), and buffing pads.

### Air Chisels

Often times during disassembly, fasteners simply will not come apart. Whether they are rusty, have rounded heads, or you simply cannot find the correct size wrench, an air chisel will get them loose. Of course, this is a last resort method, as the fastener will be destroyed in the process. However, if it fits in the first two of the aforementioned categories, it should be replaced anyway.

An air chisel works just like its predecessor, the hammer and chisel, but with a lot less effort. Air chisels require lots of air volume to work properly, but any time you are trying to remove a bolt that has a rounded head, you will be glad to

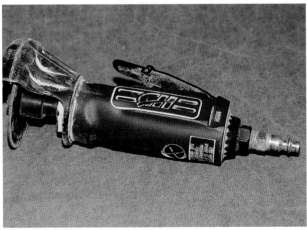

This is a pneumatic die grinder that can be used to make straight cuts in metal. It is typically used as a cut-off tool for small pieces of material or to cut out old metal where a patch panel will be installed. Prices range from less than $50 up to about $100.

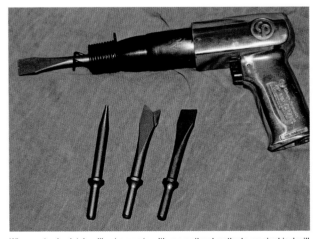

When parts absolutely will not separate with conventional methods, an air chisel will most likely do the job for you. Whether rivets or bolts used to fasten components together are rusty, cross-threaded, or just plain stubborn, an air chisel can make quick work of removing them.

have one of these at your disposal. If your shop works with lots of vintage chassis that are held together with rivets or if you need to cut apart a metal body quickly, an air chisel will be of great value.

## Smoothing Tools

Depending on the grit of the media that you are using, what you are using it on, and how you are using it, a smoothing tool may indeed be used for smoothing, or it may be used to roughen the surface, or perhaps even clean the surface—somewhat confusing I suppose, so let me explain. Media blasters are typically used to remove rust or other contaminants from the work surface, thereby cleaning it prior to other work being done. Grinders can be used for the same purpose but are also used to smooth welds and for various other chores. Sanders are pretty much limited to sanding and smoothing.

### *Grinders/Sanders*

Orbital grinders and sanders are very similar in design and operation, with their differences mainly being what type of accessory is being used with them at the time. Additionally, grinders typically run at faster speeds than sanders. Pneumatic and electric-powered models are available.

Grinders are typically used with a rigid abrasive wheel or with various types of wire wheel attachments. When equipped with an abrasive wheel, a grinder is typically used as a cut-off wheel to cut structural steel to length, clean up welds, or remove rust from heavy metal such as chassis components. When equipped with a wire wheel, a grinder works well for removing paint or light surface rust from heavy metal, or structural steel.

Orbital sanders are typically fitted with a semi-rigid pad and can be used for sanding or for buffing. With 36- to 80-grit sanding discs, a sander can be used to remove old body filler or paint from sheet metal or for light cleanup of welds, thus preparing the surface for new applications of filler or primer. You should not attempt to remove paint from an entire vehicle with a sander. You would no doubt create enough heat to cause significant warping, you would go through several sanding discs, and this process would take too much time. Even when stripping a small area, you should move the sanding disc to prevent heat buildup. You should also hold the sander so that the sanding disc flexes slightly, rather than being flat against the surfaces being stripped. Setting it flat causes the disc to jump around unpredictably, whereas applying only a portion of the disc to the metal keeps the amount and direction of the force you must resist in holding the tool fairly consistent.

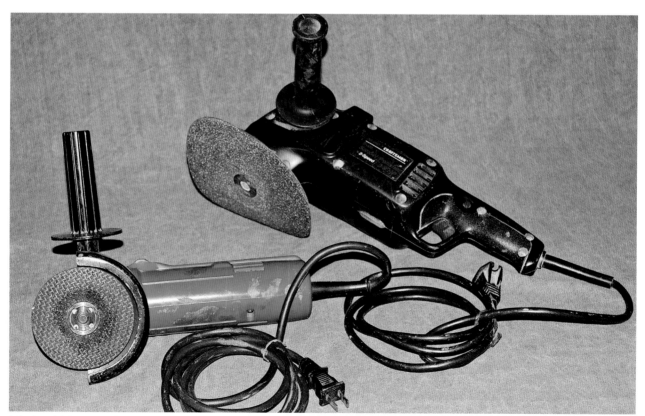

At the lower left is a 4 1/2-inch Makita grinder that can use a rigid grinding disc for grinding welds. It will also accept a wire cup brush, which is useful for removing old body filler or paint. The larger unit is a 7-inch Craftsman sander/buffer. With sanding discs of various grits, it can be used to clean welds or remove old paint or body filler. It can also be fitted with a buffing bonnet to apply wax to finished paint.

For lightly scuffing bare metal to achieve proper adhesion of primer or knocking off the initial rough edges of body filler, a pneumatic sander like this, fitted with the appropriate grit sandpaper, will quickly get the work done.

Two other types of sanders are also commonly used for bodywork: DA (dual action) and inline sanders. The DA has long been a staple in most any body shop for smoothing body filler on curved or rounded surfaces. Dual action sanders use round sanding discs similar to those used for orbital sanders. The difference is that as the sanding pad rotates, it also oscillates in an effort to smooth the overall surface, rather than simply smoothing in just one location. Still, you must manually move the DA to smooth an entire panel, as you would any other sanding device.

Inline sanders work in much the same way as a DA, but they are simply moved back and forth along one axis. Being relatively long and narrow, they are ideal for smoothing bodywork on surfaces that have no curve or curve in only one direction. (In other words, not on compound curved areas.) When working on a surface with a slight curve, orient the inline sander so that its axis of movement is parallel with the flat shape and manually move the sander perpendicular to this to work across the curved direction.

### Blast Cabinets

A blast cabinet is nothing more than an enclosed space where you can blast off rust and scale without filling the whole shop with dust. The cabinet is an enclosed box that contains a pressurized air nozzle through which blasting media is sprayed at an object that is maneuvered by hand through access holes in the front or side of the cabinet. A transparent panel in the front or top of the cabinet allows you to see what you're doing inside.

The benefit of a blasting cabinet is also its drawback. Because it's an enclosed space, you can only use it for objects small enough to fit inside and still allow the cabinet to be fully closed. Small cabinets with a working capacity of about 2x2x3 feet can be purchased for around $300. Larger cabinets that will hold objects approximately 5 feet long are available in the $1,500 to $2,000 range. Several cabinets of intermediate sizes are available and are priced accordingly.

For cleaning relatively small parts, an enclosed blasting cabinet will keep the blasting media contained. A viewing panel in the top allows operators to see what they are doing. Sleeves with gloves are attached to two openings in the front of the cabinet to allow operator access while blasting.

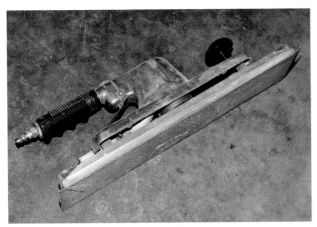

Pneumatic sanding boards that move in a straight-line direction are very useful for block sanding body filler to get it flat. However, they should not be used for removing paint or surface rust; an orbital sander/grinder is more effective.

An air compressor is used to push the blasting media through the nozzle with sufficient force to remove previous finishes. The type of finish and the material to which it is adhered will determine which type of media should be used. When purchasing an air compressor to be used for media blasting, do yourself a favor and choose one that provides more pressure and volume than you think you need. Too little pressure is a common cause for unsatisfactory results.

The way designers overcame the dilemma of creating a closed box that remains sealed even while you reach into it is to attach two rubber gloves to the access holes. The gloves stay in the blasting cabinet and you reach into them to wield the media spray gun and the part you're working on. In addition to keeping the pressurized blasting media inside the cabinet, the heavy rubber gloves also stop it from doing to your skin what it does to rust or paint.

A feature of better blasting cabinets is an integral dust collector, although a separate shop vacuum can be used. Most commercially available blasting cabinets are designed like a hopper so that blasting material automatically flows to an access point, where it can be dumped into a container, strained, and reused.

### Media Blasters

Whether you are removing rust from some heavy metal such as an automobile chassis or lighter material such as a steel fender, a media blaster is an effective and economical way to do it. However, you must use some prudence when using any type of media to blast parts, or you will quickly ruin them.

Sand blasting is a type of media blasting that has been around a long time. Although some sand blasting is still done, mostly as cleanup of steel bridges and masonry structures, there are better blasting media for use on automobile projects. Sand is very abrasive and can warp and damage automotive parts. Media that are more appropriate for automotive use include glass beads, walnut shells, plastic beads, aluminum shot, aluminum oxide, and silicon carbide.

### Spray Guns for Applying Automotive Primer/Paint

For applying primer and paint products, the most efficient and practical method is with a spray gun. Various specialty paint products are sold in spray cans only, but for new construction or refinishing projects, a spray gun is the best way to go.

Spray guns for applying automotive paint fall into one of three categories and are of one of two basic designs. The categories are full-size, touchup (a.k.a. doorjamb), or airbrush, while the designs are suction (or siphon) feed or gravity feed. There are also pressure pot systems, but these are more suited to production line use and, due to the advent of robotic spraying equipment, are seldom used in automotive applications.

### Full-Size

With a cup that will hold approximately one quart of sprayable material, full-size spray guns are the most efficient method for applying primer and paint to large surfaces and in most any situation that does not involve restricted access. Siphon feed guns have been used for years; however, stricter air pollution laws led to the development of HVLP (high volume, low pressure) spray guns that are gravity feed. This design causes less paint overspray—more paint stays on the panel, less bounces off, and the surrounding air stays cleaner.

At left is a conventional siphon feed spray gun, while the one on the right is an HVLP (high volume, low pressure) type. Purchased new for around $100 each, these are on the inexpensive end of the spectrum. Unless you are painting vehicles on a daily basis, these would probably suit your purposes.

Although many of these full-size spray guns are advertised to be all-purpose, how efficiently they operate depends on the size of the spray gun's fluid tip and the material being sprayed. Primers and other substrates are typically thicker than top coats (paint and clear), while heavy coatings such as bed liner material are thicker still. These thick materials require a fluid tip of about 2.2mm in size, while primer can be sprayed through a fluid tip that is between 1.5 and 1.8mm. Base coats and clear can usually be sprayed with fluid tips between 1.2 and 1.5mm in size.

When purchasing a spray gun, you should verify that it includes a fluid tip that is compatible with what you will be spraying. Most spray guns have removable fluid tips but are sold with just one size tip included. However, some spray guns include different size fluid tips. If you are going to be doing lots of bodywork and painting, you would probably be better off purchasing one gun for spraying primer and another for spraying top coats.

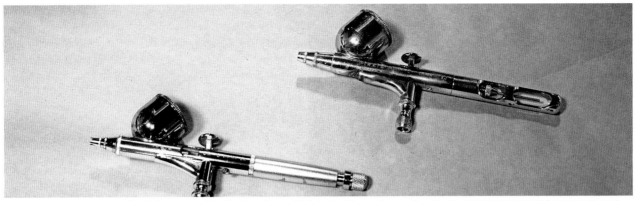

These two airbrushes are not used to paint large areas, but rather to apply very detailed artwork to any size area. The airbrush at the top is an Iwata HPC model and is used to apply relatively large areas of background or fill-in color to artwork. The airbrush at the bottom is a Custom Micron and is used extensively for superfine detailed artwork. Notice the extremely small paint cup size on each.

### *Touchup Guns and Airbrush*

Having a much smaller materials cup, touchup guns are designed for making minor finish repairs that require a small amount of paint. With their smaller overall size, they are perfect for use when refinishing small areas such as doorjambs or anytime that a full-sized gun is not practical due to space limitations. Like full-size spray guns, touchup guns are available in both siphon feed and gravity feed designs.

With the precise control that is afforded to the operator of an airbrush, they are designed for producing artwork, lettering, or any other painting that requires extreme detail. Both the touchup gun and the airbrush use an overhand grip that is operated by pushing a button or lever with your index finger, while full-sized guns are operated by squeezing a trigger.

Detail spray guns are also known as jamb guns, as they are particularly suited for jobs in confined areas, such as spraying doorjambs. They can be used to spray primer or top coats. Their paint cup is much smaller than that of a full-size spray gun, so using a detail gun for painting or priming an entire panel would require you to fill the paint cup several times.

## BENCH TOOLS

This next group of equipment includes tools for cutting, shaping, and smoothing metal, but are not necessarily pneumatically operated. Some of these tools may include pneumatics to make the job easier, but air power is not a necessity for their basic operation.

When working with sheet metal, you must become accustomed to gauge (or gage) when referring to material thickness. Most data sheets for these tools and welders (which will be discussed later) speak of gauge when discussing the capacity of the equipment in question.

| Gauge | Decimal Equivalent (inch) | Gauge | Decimal Equivalent (inch) |
|---|---|---|---|
| 24 | 0.023 | 9 | 0.150 |
| 22 | 0.029 | 5/32 | 0.156 |
| 1/32 | 0.031 | 8 | 0.162 |
| 20 | 0.035 | 7 or 3/16 | 0.188 |
| 18 | 0.048 | 7/32 | 0.219 |
| 16 or 1/16 | 0.062 | 1/4 | 0.250 |
| 14 | 0.080 | 5/16 | 0.312 |
| 13 | 0.090 | 3/8 | 0.375 |
| 3/32 | 0.094 | 1/2 | 0.50 |
| 12 | 0.105 | 5/8 | 0.625 |
| 1/8 | 0.125 | 3/4 | 0.750 |
| 11 | 0.120 | 7/8 | 0.875 |
| 10 | 0.135 | 15/16 | 0.938 |

Most sheetmetal goods are measured or described by gauge (a.k.a . gage), rather than an actual thickness.

## English Wheels

English wheels are used for forming compound curves in aluminum, mild steel, copper, brass, and stainless steel sheet material. Available in various styles and sizes, English wheels can be freestanding or bench mounted. Construction will vary by manufacturer, but the major components are the upper yoke, which mounts the upper wheel; the lower yoke, which mounts the lower anvil; a jackshaft; and an adjustment wheel, which raises and lowers the lower yoke.

12-inch or 8 ½-inch, will produce a flatter curve such as for a door skin. Each lower anvil also has a relatively narrow flat surface where it aligns with the upper wheel. This should not be confused with a flat anvil wheel, which is used for finishing and smoothing the metal after the desired contour is achieved.

Metal is stretched where the high point of the lower anvil pushes the metal against the surface of the upper wheel. How much the metal stretches depends on the hardness of the metal and its thickness and the pressure applied by the lower anvil. More pressure causes more curvature, while less pressure causes less curvature. The pressure of the lower anvil

Curved panels can be made from flat sheets of steel on an English wheel. Different crown wheels on the lower position along with varying amount of pressure will provide a tighter or flatter radius.

Material is stretched as it is moved back and forth between the lower anvil (wheel) and the upper wheel. The lower anvil has any of a variety of different crown heights, that is, the anvil will have a barrel shape opposed to being cylindrical. Common crown heights are 2 ⅜-inch, 3 ¼-inch, 5-inch, 8 ½-inch, 12-inch, and flat. Using a high crowned anvil, such as 2 ⅜- or 3 ¼-inch, will produce a tighter radius curve such as for a fender, while a low crown anvil, such as

Lower anvils (wheels) have different crowns, allowing the operator to make flatter or tighter radius curves. The closer to cylindrical (low crown), the flatter the curve will be. The higher the crown, the more pronounced the curve will be in the metal being worked.

is adjusted by an adjustment wheel located at the bottom of the jackshaft. This adjustment wheel is often designed to be kicked with the operator's foot, so that pressure adjustments can be made on the fly while still keeping both hands on the metal being worked.

As the metal passes between the upper wheel and the lower anvil, a shiny track is produced on the metal where contact was made. This tracking pattern helps the operator to duplicate the shaping onto additional pieces of metal.

This is a homemade planishing hammer, which can be made rather inexpensively. The metal tab shown at the left of the tube frame is for clamping the planishing hammer in a bench vise. An air chisel with a hammer bit, clamped to the frame with a saddle clamp, works against a smooth anvil located at the opposite end of the tubular frame.

Shiny tracks on the metal shows where the upper wheel has passed. As seen in this photo, most of the work has been done closer to the operator's right hand.

An important thing to remember when producing multiple parts is that they will most likely need to be opposite each other—in other words, sheetmetal parts that come in pairs are usually mirror images of one another used on opposite sides of the car.

Since any English wheel is going to be a substantial purchase, you must consider what size will be necessary. If you are going to be working on very small parts, such as door bottom patch panels, a model that can be clamped into a vise on your workbench may be suitable. If you are going to be fabricating fenders and deck lids, a larger, freestanding English wheel with a deeper throat will be necessary.

## Planishing Hammers

The biggest difference between a planishing hammer and an English wheel is that the planishing hammer utilizes a hammerhead and forming dies in places of the English wheel's upper wheel and lower anvil, respectively. The main purpose of the planishing hammer is to remove dimples and smooth metal after you've shaped it on an English wheel or other metal shaping device. Like the upper wheel of an English wheel, the hammerhead of a planishing hammer is flat. The forming dies work like the lower anvil and have various shapes. With the metal placed between the hammerhead and the forming die, the hammer head (mechanically or pneumatically operated) presses down on the metal several times to smooth the metal to the shape of the forming die.

## Sheetmetal Shears

Sheetmetal shears are used for making straight cuts in flat sheet metal. By aligning the sheet metal with one of the side guides, you can make square cuts easily. By holding the sheet metal in the middle of the table, an indefinite number of cuts can be made to produce a sheetmetal polygon; each individual cut will still be straight.

Three design features play a major part in the relative cost of a sheet metal shear. First is the overall size of sheet metal that can be cut on the table. Second is the thickness of sheet metal that can be cut. Third is the mode of operation—manual, hydraulic, or air operated. The larger or thicker the piece of sheet metal that can be sheared on any given unit, the more expensive it will be with all other considerations being the same. Manually operated units require stepping down on a pedal bar that spans most of the width of the machine. Hydraulically and pneumatically operated shears use a foot pedal that is connected to the drive system.

## Sheetmetal Brakes

Sheetmetal brakes are used for bending flat sheet metal. Other than its use, a sheetmetal brake has the same basic characteristics of a sheetmetal shear, in that it can handle a predetermined maximum size and thickness of material. Except for large machines that are designed for heavy industrial use, most sheetmetal brakes are operated by hand.

Although sizes, capacities, and design differ from one manufacturer to another, there are two categories of brakes: straight brakes and box-and-pan brakes. Either of these will have some type of hold down mechanism to keep the material in place. When the operator moves a lever, the front bed of the brake will pivot along a horizontal axis, bending the sheet metal along a straight edge in the process. Since the brake is hand operated, virtually any amount of bend can be made up to around 120 degrees. Even 180-degree bends can be made; however, this would most likely require a two-step process.

## Straight Brakes

A straight brake is for making simple bends along parallel axes. As an example, you could easily bend a piece of sheet metal so that when viewed from the end the edge would resemble an L, an M, an N, a V, a W, or a Z.

For making uniform bends in sheet metal, a sheetmetal brake is the tool of choice. Sheet metal slides in between the horizontal bed and the removable fingers of this pan brake. After the sheet metal is positioned and clamped in place, the front bed pivots upward, bending the sheet metal in the process.

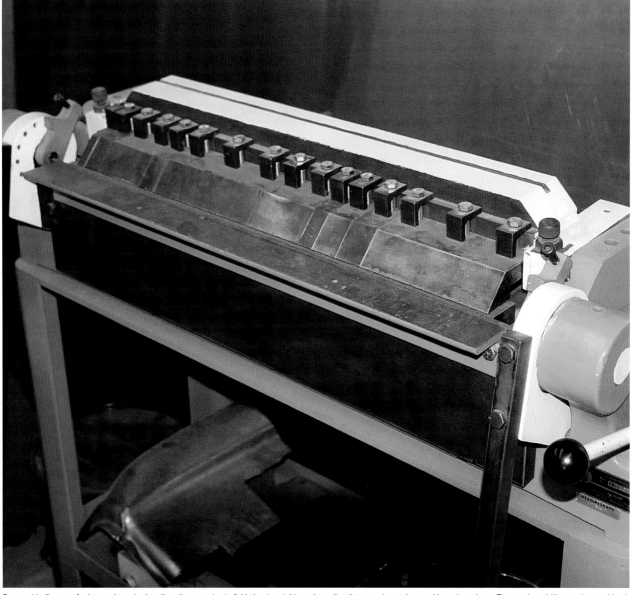

Removable fingers of a box and pan brake allow the operator to fold sheet metal in various directions, such as when making a box shape. The varying widths can be combined for bending, yet provide space in between bends along another axis.

## Box and Pan Brakes

For making complex bends along nonparallel axes, such as when making a box or a pan, a box and pan brake would be necessary, hence the name. The difference between this and a straight brake is that the edge that the metal is bent against is made of independently removable fingers of varying width. When all of the fingers are installed, the box and pan brake operates just like a straight brake. However, being able to remove one or multiple fingers allows the operator to bend portions of the sheet metal, while other portions along the same axis are not bent. The removable fingers also allow for more bending to be done as the sheet metal is bent into a three-dimensional shape.

## Bead Rollers

Bead rollers are useful for rolling strengthening beads into sheet metal. Sheet metal is fed between two mandrels that are each mounted to the end of two long cylinders. The shape of the machined mandrels is what gives the sheet metal its new shape. A bead mandrel would have a cylindrical shape with a convex ring around it near the end. The mating mandrel to this would have a concave ring that aligns with the convex ring, resulting in a raised bead in the sheet metal.

A bead roller can be used to roll beads or to flange panels, depending on the shape of the mandrels. This particular example clamps into a bench-mounted vise and is cranked by hand (rather than having a drive motor).

This is the resultant rolled bead. Note that this raised portion is what would be on the underside of the metal as it goes through the bead roller. You can therefore roll beads that protrude or recess on the finished panel, depending on which way the metal is inserted through the roller.

These mandrels are used to roll a bead in the sheet metal. The metal is inserted in between these two rollers and fed by rotating the crank at the opposite end of the roller shafts. As the sheet metal is fed through, it takes on the shape of the mandrel, in this case a rolled bead.

This shows how the sheet metal fits between the upper and lower mandrels of the bead roller. The large ring shown on this upper roller will create a bead pushed downward on the sheet metal. How the metal is run through the bead roller determines if the bead extends outward or recesses into the panel.

The shrinker squeezes the horizontal flange of this metal angle with each press, causing the metal to slightly overlap, resulting in a curved shape.

The vertical portion of the angle will remain unchanged, so this metal will eventually curve to the right. If the stretcher were being used, this piece of metal would curve to the left.

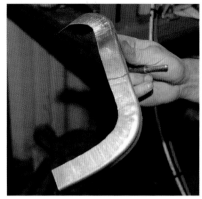

By shrinking and/or stretching both flanges of a piece of metal angle, that metal can be made to conform to multiple contours and virtually any shape.

## Shrinkers and Stretchers

Occasionally, body fabricators will have to form a piece of metal angle into a curved shape, such as for around the wheel opening of a fender, or when building a windshield frame or a headlight bezel. A shrinker/stretcher makes this an easy task. The shrinker will shrink metal on one flange of an angle to make the second flange the outside of the curve. The stretcher will stretch metal on one flange of an angle to make the second flange the inside of the curve.

## Grinders and Buffers

Bench or pedestal-mounted grinders or buffers are common in machine and fabrication shops. They utilize an electric motor with a shaft protruding from each end and a mandrel for holding grinding wheels or buffing wheels. Grinders will often have safety shields to protect the operator from flying sparks and will sometimes have a light to better enable the

operator to see the work surface. Buffers generally do not have these options so that access to the buffing wheel is less restricted. Protective shields or not, be sure to wear eye protection whenever using either of these tools.

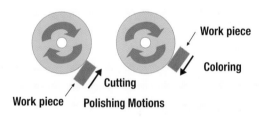

When cutting a work piece to remove surface seams, scratches, or other irregularities, move the piece in the opposite direction of the buffing wheel's motion. This is typically done by pulling the work piece toward you as the buffing wheel is rotating away from you. Cutting should be done until the surface is uniform in appearance. When buffing to bring out the color of the work piece, it should be moved in the same direction as the rotating buffing wheel, using slightly less pressure to avoid burning the work piece. This is done using various buffing wheels and compounds until the desired finish is obtained.

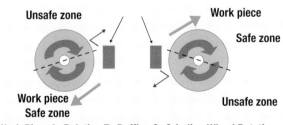

Work Piece In Relation To Buffing Or Grinding Wheel Rotation

Whether you are grinding, cutting, or buffing, you must be aware of the potential for danger when introducing a work piece to a spinning wheel. The work piece should not contact the grinding or buffing wheel on any location where the wheel is rotating toward you. The work piece should only contact the rotating wheel in an area that is rotating away from you, so that if the wheel grabs it, the object is pulled away rather than thrown at you.

Whether you build your own from a discarded electric motor and some pulleys, or buy a new buffer, you will need one for plating and polishing. The ⅓-horsepower buff motor shown is from Eastwood and is affordably priced even for a hobbyist. If you plan to get real serious about doing this type of work, you could save yourself some money in the long run by purchasing a slightly larger unit in the beginning.

## Louver Press

While many fabrication jobs can be accomplished by a variety of methods, punching louvers requires a louver press. Louvers can be large or small and can have a variety of shapes. Some have rounded corners, while others have sharp corners. The size and shape of the louver is directly determined by the size and shape of the louver die. With each operation of the louver press, the die punches through the sheet metal with a two-step process. The first step slits the opening in the metal, while the second step stretches the metal to form the shape of the louver.

The layout for louvers must be marked on the opposite side of the metal to be louvered. Like a rolled bead, louvers can extend outward or recess inward from the panel, depending on how the metal is fed through the louver press. An index mark on the louver head is aligned with the centerline, and the press is operated by a foot pedal.

## HAND TOOLS

Regardless of how well-equipped your shop may be with pneumatic and electrically operated tools, there will always be a need for basic hand tools, such as various hammers, body dollies, sanding boards, and clamps. Knowing how to use

These are some of the basic tools required for doing automotive bodywork. Beginning at the upper left is a cheese grater type of file, a slide hammer, a small sanding block, a lead file, a long sanding board, a body hammer, and a dolly.

these basic tools efficiently will make you a better fabricator. Fancy tools just make the process easier and faster.

## Hammers and Mallets

Some chassis and body building situations call for a big hammer, while others call for a smaller hammer and a lighter touch. Having both available and knowing when to use them will enhance your fabricating skills. You may need a big hammer to straighten a portion of an original car chassis, where a light hammer just wouldn't do the trick. However, using that same big hammer on an aluminum hood or steel patch panel will do more damage than good.

Most body hammers have a head and a pick, making each hammer a dual-purpose tool. The head is usually large (between 1 and 2 inches in diameter) and relatively flat with a smooth surface, while the pick end is much smaller and pointed. The larger head is used for flattening metal against a

Body hammers come in a variety of shapes, sizes, and uses. Those with a serrated head are used for shrinking metal. Round heads are used for general panel flattening, while square heads are used for restoring body lines.

dolly. The pick end is typically used for hammering out very small, localized dents, with or without a dolly. Picks can come to a very narrow point or to more of a blunt point.

Mallets are generally used to hammer flat sheet metal into a custom shape, while hammers are typically used to return damaged auto body sheet metal to a preformed shape. The striking surface of a mallet is usually made of plastic or some

Mallets are typically used for hammer forming flat sheet metal into custom shapes. They are often used in conjunction with a beater bag to form a freeform shape or against a buck or hammer form to created a specific shape.

other composite that will not mar sheet metal or aluminum. Mallets are available in different sizes and may be cylindrical or teardrop in shape.

### *Shrinking*

Shrinking hammers are similar to other body hammers, with the difference being that shrinking hammers have a serrated face instead of a smooth face. Whenever sheet metal is bent in an automobile accident, the damaged metal stretches. The serrated surface of a shrinking hammer allows a skilled user to remove some of this stretch when straightening metal.

### *Beater Bags*

Beater bags are usually made of leather or other similar material and are filled with birdshot or other ball-shaped metal. They are used for freeform shaping of sheet metal or aluminum with a mallet. Unlike a body dolly, which would cause a distinct shape in the back of the metal struck by a hammer, a beater bag's loose filling allows it to "give" somewhat.

### *Hammer Forms*

A mallet is usually used to perform hammer forming, but that is beside the point. The hammer (or mallet) form is usually made of hardwood and is shaped to the specific shape that the metal is to be hammered into. With lots of time, practice, and finesse, sheet metal can be hammered into virtually any shape by using a hammer form. Hammer forms are useful for rounding a lip onto an opening in sheet metal or creating a uniform but freeform shape.

## Dollies

Made of hardened steel that has been smoothed, dollies come in a variety of shapes and sizes. Dollies are usually held on the back side of the metal being straightened, while a hammer on the outside flattens the metal between the two tools, resulting in metal that is roughly the shape of the portion of the

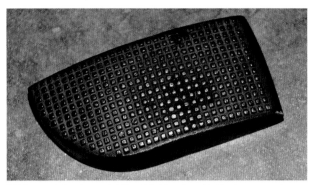

With its serrated surface, a shrinking dolly (shown) or shrinking hammer can be used to shrink metal that is stretched in a collision. Hammering the metal against or with the serrated surface causes the material to bunch up, thereby shrinking in the process.

dolly being used. For this reason, having a variety of dollies with small, large, convex, and concave shapes will give you some versatility.

## Sanding Boards

Large, small, narrow, wide, flexible, or stiff, you should always use a sanding board or block when sanding body panels, or you are wasting your time. The palm of your hand has soft spots and it has hard spots (knuckles), which put uneven amounts of pressure on the back of the sandpaper. Rather than flattening imperfections, this uneven pressure helps to accentuate waves in sheet metal, which is one of the biggest impediments to a high-quality paint job.

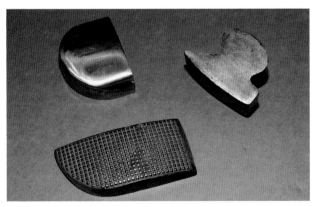

Just like hammers, dollies come in a variety of shapes and sizes. A wider variety of shapes will enable you to re-create original body lines more easily. At the upper left is a toe dolly with increasing and decreasing radii; at right is a general purpose dolly with a variety of shapes; and at the bottom is a serrated dolly for shrinking metal.

These are just a few of the many types and sizes of sanding boards. The three Flexsand boards are in about the middle of the road in terms of stiffness; they secure the sandpaper with hook-and-loop backing. The two sanding boards with hardwood handles are the stiffest and therefore are best for large flat surfaces. They secure the sandpaper with a spring clip. The small rubber block at the lower right is the most flexible. Sandpaper fits into a slot on each end and is secured with a couple of tacks that are part of the block.

## Clamps

Any time you are planning to fasten multiple pieces of metal together, with welding, screws, rivets, or other fasteners, you will need clamps to hold the metal in place. The more securely and precisely you can hold the metal pieces to be joined, the better the result you'll achieve. Spring clamps, bar clamps, magnetic clamps, and countless others can be found in most fabrication shops.

# PORTABLE STANDALONE TOOLS

These are those tools that can be easily moved around the shop, but are beyond anything that you can easily load up and take with you on a routine basis. That doesn't mean that you cannot take these tools with you on location if necessary.

Most any hot rod fabrication shop will have a gas welding setup somewhere, even though it may not be used often. Depending on how it is used, a gas welder can cut metal apart or weld it together and is often just the ticket for quick and dirty tasks.

## Welders

The ability to securely attach multiple pieces of steel together is an invaluable skill to almost anyone, but especially a hot rod builder. Hot rod fabricators use three basic welding types: gas, MIG, and TIG.

### Gas

Gas (a.k.a. oxyacetylene) welding is the oldest and the least expensive of the welding processes used in hot rod fabrication. Oxyacetylene welding, which uses oxygen and acetylene in equal pressures, remains a suitable form of welding and is still largely used in the aircraft industry. The inner core flame temperature is approximately 5,000 degrees Fahrenheit, so it is comparable to TIG welding in that respect. However, the temperature at the outer flame is much less.

The biggest disadvantage of gas welding is that the heat created by the welding process is spread over a much larger area than with MIG or TIG welding, increasing the risk that the welding process will warp or distort sheet metal. For structural steel or nonornamental applications, this is not a problem and a gas welder will be a good choice. Innumerable hot rods have been built using gas welding. However, now there are better methods available.

### MIG

Metal inert gas welding (a.k.a. wirefeed) is probably as common in contemporary hot rod builders' shops as gas welding setups were during the 1950s. MIG welding requires a power source, a wire-feeding unit, and shielding gas.

Available in sizes that run on ordinary household current (115/120-Volt AC), MIG welders are becoming very commonplace in most every fabrication shop. Relatively easy to learn, MIG welding is quite suitable for installing patch panels in sheet metal. Larger units requiring 230-Volt AC can be used for chassis quality work. Welding wire is fed by squeezing the trigger, with the wire and work surface being heated simultaneously.

The welder itself typically works as both the power source and the wire-feeding unit, as long as it is plugged into the correct voltage electrical service. Once the arc is struck, wire is fed through the welding gun (or nozzle) as long as the trigger is squeezed (and there is wire on the spool).

Several MIG units are available that run on ordinary household current and therefore do not require installing higher-voltage outlets. These units are more than capable of welding sheetmetal patch panels in place and can typically handle material up to about ¼-inch thick. Larger units that require 230 volts alternating current are more suited to welding heavier materials, such as chassis. Of course, the capabilities differ from manufacturer to manufacturer and model to model.

A disadvantage of MIG welding is that the amps, volts, and wire-feed speed must be properly adjusted to suit the conditions and the material being welded. Since this is the typical drawback of MIG welding, at least one manufacturer has introduced a line of MIG welders that feature automatic setting capabilities. Although this doesn't necessarily teach the welder operator how to decide for him/herself, it does take much of the guesswork out of welding. It doesn't really matter if the operator or the machine decides the correct settings, as long as they are correct and the weld achieves proper penetration.

### TIG

Tungsten inert gas welding produces some of the best-looking welds, largely due to the high controllability of the process. TIG uses a very high temperature to heat the base metal, but the heat is confined to a very tight area. The ability to control the arc allows the user to weld very thin metals to thicker metals without burning through the thinner material. Although welders typically use filler rod when TIG welding, it isn't necessary. A TIG machine can perform fusion welding, with no filler rod, resulting in no seam buildup.

Unlike a MIG welder, a TIG requires the use of a foot pedal or other remote device to control the torch. In very simple terms, a TIG welder uses high temperature to heat the surface to be welded, and then welding rod is inserted into the welding puddle. A simple process to understand but difficult to master.

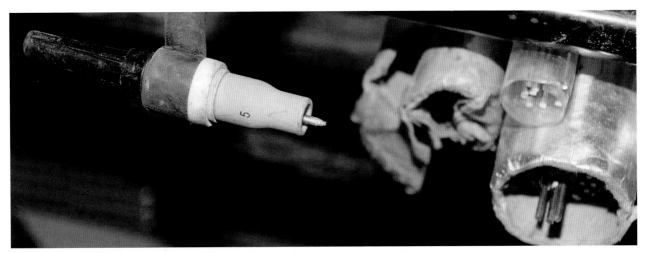

A short piece of tungsten electrode extends out of the TIG torch to transfer the heat to the surface to be welded.

Being able to take advantage of this controllability does require practice and coordination. While a MIG welder operator simply squeezes a trigger to feed the material and arc, the TIG welder operator uses a foot control, hand control, or torch control, while holding the torch in one hand and the filler rod in the other. Obtaining the correct amount of welding puddle by using one of the aforementioned controls and knowing the correct time and amount of filler rod to add is not something that just happens by accident. As with any other skill, however, you can master it with practice. Seasoned professionals make it look easy. Once you become a good welder, your fabrication capabilities become almost limitless, and you can take on any project you can dream up. For more on welding techniques, read *How to Weld* by Todd Bridigum, Motorbooks, 2008.

## Band Saws

With the ability to weld metal together comes the need to cut small pieces of metal from larger pieces so that they can be turned into brackets, gussets, patch panels, and who knows what else. For cutting pieces from bar or tube stock, or any other structural shapes, a metal-cutting band saw works great. Simply clamp the metal in the saw at the correct location to give the correct length cutoff piece, turn the saw on, and let it do its job—keeping your fingers well clear.

This smaller and less expensive band saw is typical in many smaller hot rod shops and will do most of the tasks that larger units will do. The big drawback is that it may take slightly longer; still, with a sharp blade it will usually not take long to cut anything that is going to be used in a hot rod's construction.

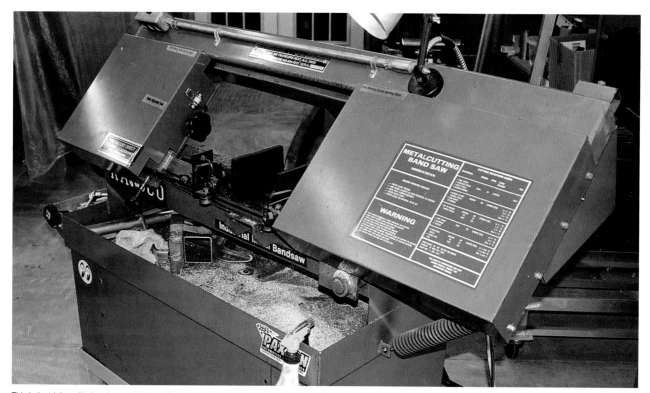

This industrial-quality band saw is typical of one that might be found in a production metal shop. With centers of the blade wheels being 6 to 8 feet apart, a fairly long saw blade is used, allowing fewer passes to make a cut than a band saw with a shorter blade. Blades on big saws therefore last longer, everything else being equal, because each tooth does less work per cut.

## Infrared Heat Lamps

Heat lamps are not required for metal fabricating, but if you use any plastic filler, or spray primer or paint, heat lamps can save time. Portable units are more suitable for speed curing small parts or localized repairs than for overall refinish tasks. However, many spray booths are equipped with infrared heaters to speed the drying and curing of overall paint jobs or to fully cure fiberglass bodies and fenders.

Although they are somewhat expensive to buy for the beginning body man, infrared heat lamps can speed up curing times for plastic filler or drying times for paint. You must make sure that you do not speed up the process too much, but if you are pressed for time, these can help you out.

## Cutting Tools

Tools used for cutting pretty much depend on what you have at hand and your overall tool budget. While a plasma arc cutter is perhaps the ideal tool for any cutting operation, their relative expense keeps them out of many shops, especially those of the hobbyist variety. Thankfully, there are alternatives that are less expensive.

Although there are better, more precise methods of cutting steel available today, a cutting torch is still very useful for removing rivet heads from vintage chassis.

### Torches

Cutting torches have been around since not long after man invented fire (give or take a few millennia). These are especially useful for cutting off rivet heads from existing chassis crossmembers and frame rails. Providing somewhat of a rough cut, they are more suitable for initial separation of bulky components that will be discarded or replaced to obtain better access to the components that are going to be resurrected.

### Bladed Tools

Other cutting tools include abrasive wheels or discs. These are generally used on sanders, grinders, die grinders, or even rotary drills as methods to cut off material from bar stock. For cutting sheet material, serrated tools, such as hand shears, tin snips, and nibblers, will do the trick, as long as the metal is not too thick.

These are compound leverage snips, which are commonly known as aviation snips. They are used to cut freeform shapes in light metal. The green-handled pair is designed to cut straight or toward the right, while the red-handled pair is designed to cut straight or toward the left. The yellow pair in the middle is designed to cut straight.

### Work Stands and Body Dollies

Work stands and body dollies can be purchased from auto body supply shops or made from a variety of scrap materials found throughout the shop. Work stands are very useful to get that fender, door, or whatever up off the floor and to a more comfortable height for working on it. Although work stands may be a luxury or convenience, they are certainly among the items that minimize fatigue, aches, and pains. The easier it is on your body to do the work, the more work you'll do, and the more you'll enjoy it.

Body dollies or chassis dollies are useful for storing and moving these large components before they are built up enough to be movable on their own. Even when you've got your rolling chassis, the body will be on and off several times as you adjust and paint, during which times you may need to move it around.

# Chapter 2
# Chassis Work

Yes, perhaps it is an overused expression, but the vehicle's chassis truly is the foundation of any hot rod project. As such, the chassis should be solid, dimensionally square, and of adequate strength to support the body, drivetrain, suspension, and payload (people or otherwise). Whether you are beginning your hot rod project with a vintage chassis, a contemporary reproduction assembly, or are building your own custom chassis, your excitement is bound to be growing intense. However, before you get carried away, to reach your ultimate goal of having a finished hot rod in the least amount of time and with little or no rework, you must do a fair amount of planning.

The suspension, as an extension of the chassis, goes a long way toward defining the style of any hot rod. Is it your desire to build a nostalgic hot rod with a dropped axle and split wishbone type of suspension? With a state-of-the-art fully independent, air ride suspension? Or something in between these two extremes? As an example, you most likely would not install an airbag suspension on a '32 Ford roadster that you are building to look like it could have been built in the 1950s. Likewise, you probably would not install a split wishbone type of suspension on a '48 Cadillac sedan. This does not mean that you cannot combine components from each of these approaches into one hot rod, but you should give the chassis plenty of thought and consideration before you start cutting, hammering, and welding.

Major suspension components and their related mounting brackets will be different for each style and will also vary from one manufacturer to another. To ensure compatibility throughout the project, it is a good practice to purchase suspension components from one source. This is not so much a hard and fast rule as it is practical advice. If you are good at fabricating brackets and adapting components, you can have success while mixing pieces and parts. However, for the average hot rodder, it makes good sense to use what an increasingly growing aftermarket has already designed and tested. Simply put, there is no need to work harder, when working smarter will serve the same purpose.

## REPAIRING AN EXISTING CHASSIS

The differences and similarities of various suspension types and components will be discussed soon, but before any suspension components are attached to the chassis, the perimeter frame should be addressed. The perimeter frame consists of the two main frame rails that run front to back of the vehicle and the crossmembers that connect the two frame rails to each other.

If you purchased a vintage automobile, the body being bolted to the frame may have kept you from adequately assessing the frame's integrity. Sure, you probably realized that you were going to pull out the original suspension. You may have given the frame some thought when you discovered that there was apparent minor collision damage to the sheet metal, but quickly forgot about it when you realized that a replacement fender would be a better choice than repairing the original. If the original frame is in good condition, you are ahead of the game, but you must verify that it is good before proceeding.

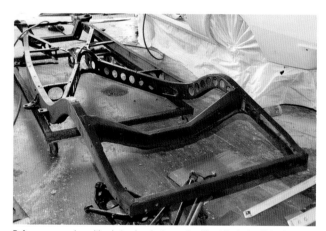

Before commencing with a hot rod project, you should get the body off the frame so you can thoroughly inspect the frame. You must verify that it is dimensionally square, not twisted, and does not have rust or other damage that would make it unusable.

Frame damage may include cracked, rusted, bent, or twisted members and resulting distortion. Depending on your resources, time frame, and budget, almost any frame damage can be repaired, but sometimes it simply is not feasible to do so. Numerous chassis shops exist solely because a repro chassis is money well spent and relatively minimal in the overall cost of building a hot rod. While many a vintage body has more patch panels and labor hours included in their resurrection than their owners care to admit, resurrecting a severely damaged chassis requires chassis straightening equipment that is beyond the hobbyist.

When it comes to rust on a frame, you must distinguish between surface rust and potential rust through. Since frame components are thicker than body sheet metal, it takes longer for the elements to eat through, which is a good thing.

So, if the frame is rusty, but fairly smooth, a trip to the media blaster or a day with a wire wheel on a grinder can make it look pretty good. However, if the frame shows significant pitting, strength may be sacrificed, making a different frame a better alternative.

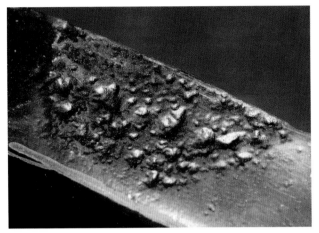

This is pitted steel, which is typically found on a vintage chassis. Although it looks somewhat rough, there does not appear to be any rust through, with even the deepest pits leaving fairly thick metal. As long as it is not rusted through, this can be cleaned up with a grinder and a skim coat of filler to look as good as new.

## Frame Inspection

You should inspect the frame carefully before you weld on any brackets, do any cosmetic bodywork, or begin painting. Although many hot rods have been built in the past without the body coming off the frame, the body really should be removed so you can do a thorough inspection. You do not need to remove any existing paint in order to measure the frame, but you should be on the lookout for any previous repairs that may need to be reworked.

Any suspension components that are not going to be used should be removed for better accessibility. Even if you are going to be using the existing suspension components, remove what you can, temporarily. With everything that can be removed out of the way, look for any obvious imperfections, such as unnecessary holes that should be filled, damaged frame rail flanges (bent, torn, or cracked), or severe rust. Do not be afraid to circle any findings with chalk or marker so that you remember to tend to them later. Be thorough, because you've got full access. Now is the time to make the chassis as perfect as you want it to be, not later when it's all built up.

### Measuring

Once you have inspected the frame for obvious imperfections, you'll need to measure it for twist and squareness. You want to build from a frame that has no twist and is dimensionally square. Begin by placing the bare perimeter frame on three jack stands—one located at the rear of each frame rail and the third under the center point of the front crossmember. (Don't use four, as that will prevent the frame from relaxing into its natural position.) Place a carpenter's level across the two rear frame rails and then use shims to make the rear of the frame level on the jack stands. The shims will compensate for a shop (garage) floor that is not level, or for jack stands that are not perfectly equal height.

With the rear of the frame level, check for level across the frame rails at the front and rear crossmembers, as well as several locations in between. To maintain consistent readings, keep the level facing the same direction. While progressively increasing out-of-level readings indicate a twisted frame, localized out-of-level readings usually indicate bumps or sags.

To determine if the frame is dimensionally square, measure diagonally from a known point (e.g., a front suspension mounting point) on the top front of one frame rail to a known point on the top rear of the opposite

Sitting out in a field, this vintage Ford frame may look to be in great shape, but, before you spring for the long green to purchase something like this, you should inspect the frame as carefully as possible. Besides checking for rust, cracks, rips, tears, and dents in the metal, you should check to see that the frame is dimensionally square.

frame rail, and record the measurement. Now measure from the similar points on the opposite frame rails, and record the measurements. Ideally, the measurements will be the same. If not, the frame is out of square. Repeat this process between the front crossmember and a point near the middle of the frame, between the middle of the frame and the rear crossmember, and between the front and rear crossmembers. Any differences between similar measurements indicate that the frame is diamond shaped, rather than square, thus needing repair. A frame may have an overall diamond shape, or localized out-of-square measurements if the frame was impacted from one side only.

In addition to checking for twist and squareness, you should verify that the distance between the frame rails is correct. Incorrect widths between the frame rails will have adverse effects on the fitment of the radiator, hood, and fenders with the body and running boards. Most chassis shops and reproduction body manufacturers should be able to provide you with the correct dimensions of the frame that you are working on. You must also verify that the frame rails are vertical. If the frame rails are not vertical, problems will develop when mounting fenders and running boards.

## Straightening

If your frame measurements revealed any imperfections or inconsistencies, you'll need to do some straightening work before proceeding. If frame rail flanges are slightly tweaked, they can usually be hammered back into shape with a hammer and dolly. However, most other frame straightening activities should be done by a competent chassis alignment shop, as the required pushing and pulling power is usually beyond the capabilities of any shop that does not perform this kind of work on a regular basis.

### Frame Rack

The basic process for straightening a frame is to place it on the frame rack, take measurements at specific locations with precise measuring equipment, and then pull the frame back into correct alignment by using hydraulic jacks that are part of the frame straightening table. A skilled operator and precise measuring equipment are essential to straighten a frame properly. This frame straightening process squares up the chassis and should not be confused with a wheel alignment, which aligns the wheels to the chassis.

## Boxing

To those who are familiar with reproduction hot rod chassis only, you may not realize that original frame rails for American cars built prior to 1948 were merely a three-sided channel, rather than rectangular tubing. When early hot rodders began installing larger and heavier engines between the stock rails of their stripped down hot rods, they quickly realized that a rectangular cross-section made a stronger frame rail than the stock channel. By cutting flat steel plate that is similar

While an unboxed frame may be strong enough to support the body for which it was designed, larger engines and the addition of independent front suspension call for at least partial boxing. Boxing plates for this particular frame, included with the front suspension kit, substantially strengthen the stock open frame rails.

By boxing only the critical portions of the frame, fuel lines, brake lines, and wiring can be placed within the confines of the stock channel, keeping them out of harm's way in the process.

in thickness to the frame rails and then welding the plate across the flanges of the stock rails, the frame becomes more rigid. Additionally, boxing the stock frame rails makes attaching mounting brackets for suspension and other components simpler.

Boxing plates can be installed the full length of the frame rails or in localized areas. On partially boxed frame rails, the typical locations are at the motor mounts primarily and in the location of the rear suspension brackets as a secondary location. In either area, boxing plates should be a minimum of 12 inches in length.

## Making Boxing Plates

For some chassis, boxing plates are commercially available. If not, you can certainly make your own. Before you start cutting out steel plate, you will need to make a pattern, as most frame rails are not straight, nor do they have a uniform cross-section. Prior to making a pattern, you must verify that the upper and lower flanges of the existing frame rail are the same width. If they are not, you will need to trim the wider rail to the necessary width, or the boxing plate will not be vertical when installed. Having nonvertical boxing plates will create problems when installing engine mounts or other brackets later. To make a pattern, you can use poster board, cardboard, paper, or even empty cereal boxes. Ideally, you will have something large enough in one piece, but you can tape smaller pieces together to gain the necessary length if need be.

One method of making a pattern for a boxing plate is to use a roll of masking paper, in a width that is wide enough to cover the widest section of the frame. Tear a piece of masking paper off the roll that is long enough to cover the length of the frame's rail (or at least the section you desire to box). Tape one end of the masking paper to the frame rail slightly beyond where you want the pattern to begin. Pull the masking paper taut against the inside of the frame rail to a point beyond the end of where the pattern should end. Now tape the masking paper to the frame rail at this end. You now have a couple of choices of how to proceed.

If you want the boxing plate to cover the inside flanges of the existing frame rail, fold the masking paper over the top and bottom flanges, making a sharp crease. This crease is where you would cut the pattern, prior to laying it out on the boxing plate material. If your preference is for the boxing plate to fit between the top and bottom flanges of the existing frame rail, use a pencil or marker to rub against the edge of the frame rail flange with the masking paper between the two. If done properly, this will leave marks on the masking paper the width of the frame rail flanges. The marks indicating the lower edge of the top flange and the upper edge of the lower flange is where you would cut the pattern.

For boxing plate patterns that will be used on a regular basis, it would be wise to transfer the original pattern onto a piece of sheet metal so that it will last indefinitely. You could then apply machinist's dye to the boxing plate material, then scribe around the pattern, leaving a scribed line in the machinist's dye. If the pattern is going to be used on just one project, secure the boxing plate pattern to a piece of appropriately sized steel plate that is similar in thickness to the existing frame rails. Then use a permanent marker to trace around the pattern.

How you will cut out your boxing plates depends on the equipment you have at hand. A plasma arc cutter would be most precise; however, band saws, cutting torches, and other methods have been used. If you take that extra bit of time to cut the plate precisely, you'll save time and get a better-looking result when you weld it in.

## Installing Boxing Plates

To install boxing plates, you will need to use as many clamps as you can muster, along with a welder. Test fit the boxing plate to verify that it fits its designated location properly and needs no additional grinding or trimming. If it needs grinding or trimming, do it now. When the boxing plate fits properly, clamp it in place. Using your choice of welding equipment, tack weld the boxing plate to the frame, with tack welds approximately every 6 inches. With the boxing plate tack welded securely, double-check for proper fit and alignment. If the boxing plate is not aligned properly, break or cut the tack welds with a grinder and adjust accordingly. If the boxing plate is aligned and fits properly, finish welding it in place.

Since they are usually only about ⅛-inch thick, boxing plates do not offer a very good base for drilling and tapping a hole and fitting mounting brackets—and with the frame rail closed off, it's impractical to fit a loose nut to the back side. An easy fix is to weld a thick plate to the inside of the outer frame rail or the boxing plate as required for the particular bracket. This plate can then be drilled and tapped to anchor your bracket. For some situations, you can weld a threaded bung to the inside of the frame instead.

## Filling Unnecessary Holes

When resurrecting an original frame or even with some reproduction frames, you will most likely have some existing holes that you do not need or want. A common cause for such holes are original rivets that have been removed during previous chassis modifications. Other holes may be from a variety of accessory installations and therefore may vary from small to large in size. Still other holes or tears may be the results of traffic accidents.

### Removing Rivets

Rivets, commonly used to secure stock crossmembers, can be removed by a few different methods. The quick and probably least frustrating method is with a pneumatic air chisel. With an ample air supply and a decent chisel head placed between the head of the rivet and the face of the frame, rivet heads can be removed quickly. Theoretically, the rivet should simply fall out once the head is removed. In reality, however, a smack or two with a hammer and punch may be required. If you do not have access to a pneumatic air chisel, you can resort to a hammer and chisel to achieve the same results, albeit with additional effort. Another approach is to remove the head by grinding if off with a grinder or drilling through it with a drill bit. Again, once the rivet head is removed, the rivet itself can be removed with a hammer and punch.

*Welding Holes Shut*

Unwanted holes in frame rails can be closed up by one of two different methods. The method chosen depends on the size of the hole. If the existing hole is approximately ¼ inch in diameter or smaller, it can simply be welded shut by building up weld around the outer edges of the hole until the hole no longer exists. When the hole is filled, grind the weld smooth.

If the hole to be filled is larger than ¼ inch in diameter, a plug should be welded in place. Begin by cutting a slice from round bar stock that is just slightly smaller in diameter than the hole to be filled. If you do not have bar stock in the correct diameter, you can drill out the existing hole to just slightly larger than the diameter of the round bar stock that you have available. Cut the slice just slightly thinner than the frame rail that surrounds the hole being filled.

Once you have the plug cut out, bevel the edges slightly with a grinding wheel or file, to allow for an appropriate weld bead. Now push the plug into the hole. If it will stay in place on its own, tack weld it in place and then weld around the perimeter of the plug. If the plug will not stay in place prior to welding, try holding it in place with a magnet or by lightly tack welding a thin piece of bar or tube stock to the outside of the plug to serve as a handle. Be sure to use appropriate welding gloves. Hold the plug in place, and weld around the perimeter of the plug. Then grind away the tack weld securing the improvised handle to the plug.

When all extra holes are welded shut or are filled with a plug, grind off any excess weld material or plug that protrudes from the face of the frame rail, so that the surface is flat. Any low spots can be filled later with a skim coat of body filler as desired.

## Replacing Crossmembers

Prior to replacing or installing crossmembers, you must position the frame rails the correct distance apart and in the same horizontal plane. If you are modifying an existing chassis, this is easy to do by supporting the frame on three jack stands as discussed previously. If you are installing crossmembers between two new frame rails, use a chassis jig to ensure the rails are square and not twisted before you proceed. (Using a chassis jig to build a new chassis is discussed in more detail later in this chapter.)

If you are working with a perimeter frame that has already passed the tests for being square and without twist, use pieces of scrap metal to maintain that condition prior to removing crossmembers. Clamp square or rectangular tubing across the bottom and/or top of both frame rails to serve as an interim crossmember. Flat bar stock should be avoided for use as an interim crossmember, as it can flex. A better alternative to flat bar stock would be angle stock if square tubing is not available. Place the interim crossmembers as close as practical to the crossmember being replaced, but not so close as to inhibit removal of the old crossmember and installation of the new one.

Remove any rivets that secure the existing crossmember with the methods described previously. If a crossmember to be removed is welded in place, cut it out with the equipment available (plasma arc cutter, air saw, cutting torch, etc.), and then prep the area accordingly with a grinder or sander so that any remnants of the existing crossmember are completely removed. If the crossmember to be removed is held in place with bolts, simply remove them by using the appropriate size wrenches. If existing hardware is rusty and difficult to remove, liberally apply WD-40 or other similar penetrating oil, and then try again. If this does not work, use an air chisel or cutting torch as necessary.

Once the existing crossmember is loose from the frame rails, it can be removed. You may need to use a hammer to move the crossmember to a point where you can pull it free. Be sure to support the crossmember adequately so that it does not simply fall out of the frame, possibly causing personal injury.

Now that the existing crossmember is out of the way, you can install the new one. Be sure to read and follow any instructions that may be included with the crossmember if it is a purchased part. Position the new crossmember and clamp it in place. Measure and double-check those measurements to verify that the crossmember is mounted squarely between the frame rails. If it is not, adjust accordingly. When the crossmember is squarely mounted and in the proper location, tack weld it in several locations, then measure again to verify that it is still in the correct position. It is much easier to cut or break the tack welds and reposition the crossmember now than it is later when it is fully welded in place. After verifying that the crossmember is in the correct location and square with the rest of the chassis, finish welding it in place. Use short welds (rather than long continuous welds) and skip around from one side of the chassis to the other to avoid building up too much localized heat. This is more important when using a MIG welder than a TIG, but it is still a good habit to develop, no matter what type of welder you are using.

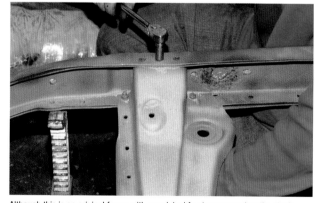

Although this is an original frame with an original front crossmember, the stock rivets had previously been removed and the crossmember bolted in place. Removing the bolts and/or rivets is the first step toward removing a stock crossmember.

Before removing the stock crossmember, clamp a piece of steel tubing to both frame rails to maintain the frame rails at the correct distance apart.

Although you can use a tape measure to determine if a frame is square, a trammel gauge is more precise. Pins that are adjustable along the gauge can slip into mounting bolt holes in the frame, making this a one-person job.

After measuring from one point on the left frame rail to a point on the right frame rail, measure to the corresponding points on the opposite frame rails to determine that the frame is square.

To prevent the frame from springing inward or outward when the original crossmember is removed, tack weld some square tubing or bar stock to the frame rails. Place this as close as practical to the crossmember that is being removed, but not so close that it is in the way.

With one temporary brace tack welded in place behind the original crossmember, another mounted forward should be used to prevent the front frame horns from expanding or contracting.

Simply clamp the tube stock in place so that it spans both frame rails, then tack weld it in place. Since this brace is temporary, it doesn't have to be precise in its placement, but if it is reasonably close to perpendicular to the frame rails, it will be less likely to induce any stress.

As another bit of peace of mind, a third temporary brace that spans the frame rails along with the stock crossmember should be more than adequate to prevent any movement or deformation of the original frame rails.

When spanning the frame rails and the stock crossmember, be sure to tack weld the brace to each to maximize the protection against deformation.

A comparison of the stock crossmember (still in the frame) and the new (albeit weathered) crossmember for the independent front suspension. While the stock crossmember fits between the frame rails, the IFS unit fits to the outside of the frame rails. The round tubes on each end of the upright portion are for mounting the upper and lower A-arms.

After removing the stock rivets (or bolts in this particular case), use a hammer to entice the stock crossmember to slide from between the stock frame rails.

Since the stock crossmember fits into the channels that make up the frame rails, it will not come out by simply moving it forward or backward. It will need to be turned horizontally to be removed. Watch your toes when you get close to having it out, as it is a heavy piece. Try to have a method of supporting the stock crossmember as it parts company with the frame, rather then just letting it fall out.

The replacement crossmember can be positioned according to the manufacturer's instructions, and then clamped in place. Double-check your measurements, tack weld it in place, and then check your measurements again. If everything is in the correct location go ahead and fully weld it in place. If the crossmember is not in the correct location, now is the time to correct it.

This is what the IFS crossmember looks like after being fully welded in place. Note the use of boxing plates on the inside of the stock frame rail to provide additional strength and to provide a convenient location for attaching weld-on motor mounts.

## Suspension Angles

Before we get to various suspension systems, let's look first at some basic suspension-related angles. While they primarily affect steering, you need to think about them when choosing suspension components so the finished vehicle rides and drives safely. Some suspension components offer a greater amount of adjustment or fine-tuning; however, if the suspension is designed properly and with some amount of forethought, it will not require adjustment. With that said, any suspension system should allow some adjustment to compensate for component wear.

### *Caster*

Caster is defined as kingpin inclination—or angle between an imaginary line between the upper and lower ball joints and true vertical (as viewed from the side of the vehicle). When the top of the kingpin "leans" toward the back of the vehicle (positive displacement of the top of the kingpin along the "x" axis), the caster is positive. When the top of the kingpin "leans" forward (negative displacement along the "x" axis), the caster is negative. Positive caster is what causes the front wheels to "self-center"—to return to straight ahead after the vehicle makes a turn. Hot rods typically have between 7 and 9 degrees of positive caster. An example of extreme positive caster is a chopper with an extended front fork forming 45 degrees or more of caster. A common example of negative caster and its ill effects is a shopping cart and the seemingly ever-present wobbly front wheels.

Caster can typically be adjusted with comparable ease, regardless of whether the front end uses an axle or an independent front suspension. A limiting factor in adjusting caster is the type of radius rod utilized on a vehicle that uses a front axle, which will be discussed a bit later. While conventional wisdom would lead one to believe that caster should be the same for both front wheels, a slight twist in the axle (new or old) will cause caster measurements to be different if the radius rods are adjusted identically. The radius rod adjustments should end up identical; however, the caster of each wheel should be set independently so that the caster is the same for both, even if this dictates that adjustments vary slightly. If the caster is not the same for both front wheels, the vehicle will tend to pull toward the side that has the least positive or most negative caster.

When designing and building the chassis, in particular, placement of the front crossmember, and the effects of tire sizes must also be considered. As an exaggerated example, suppose that the chassis was straight, having no upward or downward curvature or bend to it, and the axle centerlines (front and back) were mounted the same distance below the frame.

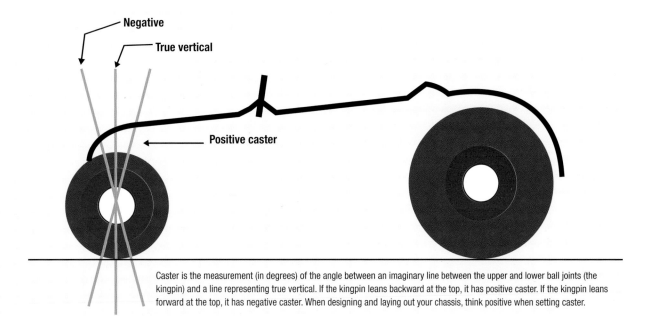

Caster is the measurement (in degrees) of the angle between an imaginary line between the upper and lower ball joints (the kingpin) and a line representing true vertical. If the kingpin leans backward at the top, it has positive caster. If the kingpin leans forward at the top, it has negative caster. When designing and laying out your chassis, think positive when setting caster.

So, if tires with the same overall diameter were mounted on the ends of both axles and the chassis was parked on a perfectly level grade, the tops of the frame rails would also be level. With this picture in your mind, imagine that the caster is set at zero degrees. In other words, the kingpin is straight up and down or perpendicular to the frame rails. Now if you replaced the rear tires with tires having a larger diameter, you have effectively caused the caster to become negative, which is bad for a hot rod. Since most hot rods have rear tires that are larger than the front tires, tire size itself will affect caster angles and typically in an adverse direction, unless you account for this when installing the front crossmember. Additionally, the shorter the wheelbase, the more pronounced the difference will be—changing to a larger diameter tire on the back of a stretched limo will produce less caster change than swapping the same larger tires onto a short wheelbase T-bucket.

## Camber

Camber is the angle (in degrees) between the vertical axis of the wheel/tire assembly and true vertical when viewed from the front or back of the vehicle. If the tops of both front tires are farther apart than the bottoms (the tires lean outward from the vehicle), the vehicle has positive camber. If the bottoms of the tires are farther apart than the tops (tires lean in), the vehicle has negative camber. While cornering may benefit from negative camber, straight line acceleration benefits from zero camber (straight up and down). However, when tire flex, engine torque, and other real world characteristics are accounted for, most hot rod suspensions utilize positive camber by design. Positive camber also requires less steering effort, making it more conducive to everyday use. Unlike caster, camber is not easily adjustable on vehicles that utilize a straight or dropped axle.

Because many stock axles have been "dropped" by heating and/or bending, a damaged axle could theoretically be repaired by unbending the damage. However, DO NOT do it this way. Some shops that specialize in repair of front axles may be able to repair a damaged axle, but you should consult with someone who can and will give you an honest answer as to the quality of their work before entrusting your axle to them. A new front axle is relatively cheap compared to the possible consequences of using one that is damaged or repaired incorrectly.

## Toe-In/Toe-Out

Commonly referred to as toe-in, toe is the angle inward or outward (from straight ahead) that the front tires point when viewed from above. If the tires point inward, they have toe-in. If the tires point outward, they have toe-out. In rear-wheel-drive vehicles, the front tires tend try to pull away from each other at their leading edges, so toe is usually set inward to compensate for this.

While toe is the measurement of an angle, it is typically measured in inches (more accurately fractions of an inch) at a specified wheel diameter. Vehicles that use bias ply tires usually have a toe-in specification of about $\frac{3}{16}$ inch, while radial tires on the same vehicle will suffice with a toe-in spec of about $\frac{1}{8}$ inch. In other words, the front of the front tires at the wheel's center tend to be set $\frac{3}{16}$ inch or $\frac{1}{8}$ inch closer together than the backs of the front tires.

Toe (in or out) is not an angle that is crucial during the chassis building stage; however, you'll want to remember to set it before you call the project done because it affects tire life. While the local alignment shop most likely will not have specs for your hot rod, if they are competent with their equipment, they should be able to align it if you provide the alignment specifications. Otherwise, the shop can probably give you what their experience suggests is right, and you can observe tire wear and adjust from there. Note that the alignment specifications suggested within this book and by parts manufacturers are merely suggestions based on past experience and are by no means complete and accurate for every hot rod.

Regardless of the type of front suspension, toe is typically adjusted by threading the tie rod ends inward or outward. If the tie rod is located behind the axle centerline, threading the tie rod ends outward will increase toe-in, while threading them inward will decrease toe-in. Conversely, if the tie rod is located in front of the axle centerline, threading the tie rod ends outward will decrease toe-in (or increase toe-out), while threading them inward will increase toe-in.

## Pinion Angle

Although pinion angle does not directly affect ride quality, handling, or steering, it must be addressed during chassis construction to prevent it from becoming a big problem later. Regardless of the type of rear axle housing or the type of suspension locating it, the rear axle pinion must be at the proper angle in relation to the output shaft of the transmission.

Ideally, the transmission output shaft and the input (pinion) shaft into the rear axle housing will be parallel, therefore having equal and opposite pinion angles. For street driven hot rods, this angle should be approximately 3 to 3 $\frac{1}{2}$ degrees; however, 6 to 6 $\frac{1}{2}$ degrees is acceptable as long as the output and input shafts are near parallel. A pinion angle greater than 6 $\frac{1}{2}$ degrees will begin eating U-joints. The farther out of these parameters that you get, the greater the risk of damaging the tailshaft bushings in the transmission and/or the pinion bearing in the rear end. While the transmission output shaft, the driveshaft, and the input (pinion) shaft into the rear axle housing must be in direct alignment when viewed from overhead, they should not be along the same line when viewed from the side. Being in direct alignment in the horizontal axis will cause the needle bearings to wear an indent into the U-joint cap, causing them to eventually stop rolling. Additionally, the engine should be positioned between the frame rails so that the base of the carburetor or fuel injection is level. So having an engine (or at least a mockup block), the rear axle housing, and the wheels and tires that will be used on the finished vehicle is essential when you install the rear suspension and axle housing.

Ideally, the rear wheels and tires that will be used on the finished vehicle are available so that they can be mounted and installed on the rear axle. The rear axle can then be rolled under the rear portion of the chassis. If the correct wheels and tires are not available, use jack stands to support the rear axle at the anticipated ride height. If the engine is already installed, use an angle finder to determine the angle of the transmission output shaft. Since mockup components are often used for this process, remember that the base of the block (where the oil pan mounts), the cylinder heads, and crankshaft are all components that are parallel to the transmission output shaft. If none of these are available or convenient, the back of the block where the transmission mounts can be used to determine an angle because this surface is perpendicular to the transmission output shaft. Thus, adjust your measurement by 90 degrees if you use the transmission mounting surface on the block. The rear axle housing can now be rotated as necessary to match the pinion angle to the angle of the transmission output shaft.

For drivetrain longevity, you want to minimize driveshaft vibration. How much a driveshaft is prone to vibrating depends on its weight, size (diameter), and overall length—though all driveshafts will begin vibrating at a particular harmonic. (We're assuming that you start with a straight driveshaft free of dents and distortion.) All other things

being equal, a lighter and shorter driveshaft with a larger diameter will be able to spin at a higher rpm level without inducing vibration than one that is heavier, longer, and smaller in diameter. In other words, larger diameter, lighter material, and shorter length help to reduce vibration, while smaller diameter, heavier material, and longer length increase vibration. For example, a driveshaft that is 60 inches long, 3 ½ inches in diameter, and made of mild steel (heavy) will begin vibrating at 4,640 rpm. The same size driveshaft made of 4130 chrome-moly will be able to spin at 5,357 rpm before it begins vibrating due to its harmonic. Using a lighter material such as aluminum will allow the same driveshaft to spin at 6,202 rpm before it begins vibrating.

By studying the diagram below, you should be able to determine the proper pinion angle setting, even if the transmission and/or rear axle third member are not installed. From high school geometry, we know that if the output and input shafts are parallel, the back of the engine block and the face of the rear axle housing must also be parallel, so the rear end can be set as long as the engine's final position is known. It is good practice to simply tack weld the rear axle brackets in place during construction and then set the final ride height with the full weight of the vehicle on the springs before doing the final welding. Likewise, the driveshaft measurement should be done with the full weight of the vehicle on the springs as well. As you go, keep a "before operating" list for your hot rod that includes all the tasks you must finish before you attempt to start and drive the vehicle. Include all final welding tasks, such as rear axle brackets, on this list in order to remove any possibility of operating the vehicle before everything is solid and secure.

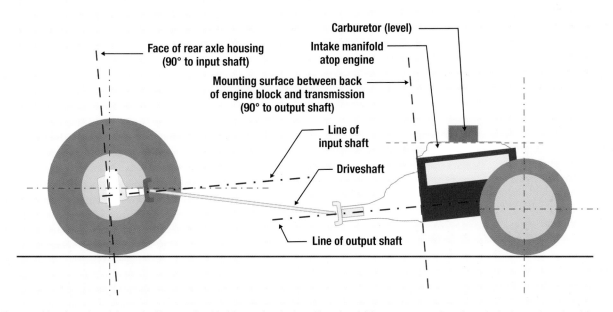

For street-driven hot rods, a pinion angle of between 3 and 3 ½ degrees from horizontal is preferred. It is not necessary to have the engine in place to determine pinion angle, but you may have to reverse engineer the motor mount location if the engine is not installed when pinion angle is established.

# FRONT SUSPENSION

The front suspension of any hot rod has two distinct tasks that it must accomplish. One is to support the front portion of the vehicle, while the second is to incorporate a steering system. Front suspensions can also be grouped into one of two basic designs: those that utilize an axle to connect the front wheel/tire assemblies to each other, or an independent front suspension that effectively disconnects the front wheel/tire assemblies from one other.

## Straight or Dropped Axle

Traditional hot rods typically utilize a straight axle or a dropped axle. As the name implies, a straight axle is one that is straight when viewed from the front or back of the vehicle. With the suspension components (springs) typically located between the axle and the frame, a straight axle usually raises the front of the vehicle. Straight axles are typically found in drag cars of the 1960s or copies thereof. Conversely, a dropped axle is bent (intentionally) so that the front of the car can be set lower than the front-wheel centerline. While original axles that were heated and bent have been dropped as much as 6 inches (perhaps more), manufactured dropped axles are typically dropped between 2 and 4 inches.

Regardless of whether you are using a straight axle or a dropped axle, it must be attached to the chassis and it must cushion the ride. Typically, there is a radius rod connected to slightly inboard of each end of the axle that runs back to the frame on each side. Common variations of the radius rod design are wishbones, hairpins, and four links. While each of these is uniquely different, they work utilizing the same basic principles.

This stock Ford Model A front suspension gives a glimpse backward at the heritage of a typical Ford hot rod front suspension. As you can see, even though many of today's components are made of stainless steel and polished or chrome-plated, there is no denying the lineage. Although this axle is not dropped, it does have a slight arc in its stock configuration. It didn't take long for early hot rodders to realize that they could lower the front of their rods by dropping the axle.

The chromed components of this hot rod make up the majority of a typical dropped axle setup. At each end of the dropped axle is a spindle, to which the wheel is mounted. A transverse spring connects to the spring perch via a spring shackle. The spring perch pins a batwing to the front axle and also has the lower shock mount threaded onto its lower end. The batwing provides front mounting points for each of the parallel radius rods. Shocks mount to the suspension at the aforementioned lower shock mount and to the frame at the top.

### Wishbone Radius Rods

In many early vehicles that became hot rods, the original axle was located by a wishbone arrangement. Bars that were connected to the axle on the front came together near the middle of the vehicle with a pivot arrangement that allowed the axle to rise on one end and lower on the opposite end as the car traveled uneven roadways. However, since the wishbones and axle were connected and in the same plane, bump steer was a common problem. As one wheel would hit a bump in the road, this would cause an equal and opposite reaction in the other front wheel, changing the tire angles against the road and causing minor or not-so-minor veering—hence the name bump steer. At the relatively low speeds when these vehicles were new, this was not a big problem. At today's highway speeds (and especially in comparison to the improved steering and handling of contemporary vehicles), stock wishbones would seem to be ill handling.

As early hot rodders began installing larger engines and transmissions between the frame rails, this stock arrangement did not provide ample room, so the wishbones were split. While the wishbones would maintain the original axle mounting locations, the rearward end of the 'bones would be connected to the frame via an adjustable rod end. Splitting the wishbones does allow for a much wider variety of drivetrain combinations when compared to the stock configuration.

While a wishbone suspension may not provide the superb handling and creature comforts of other suspension systems that are now available, it was the best thing going for a time in rodding's history. If you are building or restoring a period perfect hot rod from the early days, split wishbones are the suspension of choice. Since the wishbones are mounted directly to the front axle and secured in place by a spring perch, there is no provision for adjusting caster, making proper design setup a critical issue. One method to allow at least minimal adjustment is to incorporate a custom frame mounting bracket that offers multiple mounting locations. As the rearward end of each of the front wishbones attaches to the frame in just one location, the length of the wishbone will have a direct correlation to the amount of adjustment that is available.

This is the central pivot point, which is usually cut off of wishbones. When this pivot (along with the curved portion of the wishbone) is removed, the "split" bones can then be mounted outward on the vehicle, allowing for larger engine and transmission combinations.

This is how the stock wishbone is mounted to the front axle of a 1935 through 1940 Ford. In this configuration, the transverse leaf spring is mounted in front of the axle.

A relatively stock (albeit split) front wishbone (top) and rear wishbone (lower). The yoke of the front wishbone sandwiches the axle and would be held in place by the spring perch. The front portion of the wishbone is dropped to clear the tie rod. At the back is the frame mounting hardware. At the back of the rear wishbone is a plate that can be welded around the axle tube. At the front is an adjustable bar end and mounting tabs that would typically be welded to a tubular crossmember.

Very simple in design is the split wishbone on this rat rod—one mounting point to the frame rail and a yoke up front that straddles the front axle. The front portion of the split wishbone is then secured to the axle by a bolt that passes through the yoke, the axle, and the yoke again. Be sure to use lock nuts or lock washers, nuts, and cotter pins to secure suspension components.

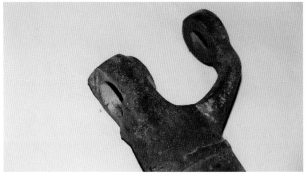

A closer look at the portion of the front wishbone that would mount to the front axle. This is typically secured to the front axle by the spring perch for a transverse leaf spring, in similar fashion to a contemporary batwing of a parallel link radius rod setup.

This is on the front wishbone, but the rear wishbone frame mount would be similar. The end of the wishbone has been cut off (split), a threaded insert will be welded on, and then an adjuster with a jam nut threaded into it. The bolt shown would actually mount to the frame rail.

### Hairpin Radius Rods

Historically speaking, the next generation of radius rods to come about were referred to as hairpins. Presumably, this was because they had a shape similar to the accessory with which women would style their hair. Although the hairpin radius rod still utilizes just one pivot at the frame mounting location, it has adjustments at each of two mounting points (one above, one below) the axle. These two independently adjustable mounting points allow for caster adjustment, making them a significant improvement over wishbones.

This particular style of hairpin radius rod attaches to the axle with the spring perch securing it to the axle. Another style of hairpin (to be seen later) attaches to a plate that is welded onto the axle.

Connecting the hairpin-style radius rod to the axle is done by one of two basic methods. One method utilizes a "batwing" that is secured to the axle by the spring perch. An adjuster threads into the hairpin and is secured by a jam nut. The adjuster then fits between the ears of the batwing and is secured by a bolt and lock nut (or lock washer and nut). This method requires no welding, as it is a bolt-together assembly of readily available parts. The second method utilizes a bracket made of steel plate that is cut to the proper shape and then welded to the axle. Adjustment is made by using a clevis threaded into the hairpin and secured by a jam nut. The opposite end of the clevis is split, allowing it to fit against each side of the axle plate. This end of the clevis is secured by a bolt and lock nut (or lock washer and nut). This method does require welding the axle bracket to the axle, unless you purchase a front suspension package that may already have the welding completed. For a competent fabricator, the axle plates would be easy enough to cut out of sheet stock, making this a low-budget method of suspending the front of a hot rod.

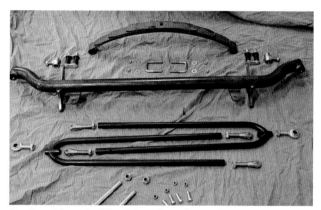

Mounting points may be different, depending on the manufacturer; however, most dropped axle front suspensions consist of similar components: a spring that is secured to the front crossmember and the axle; the axle, to which are mounted the wheels, tires, and brakes; and then radius rods that locate the suspension assembly. Not shown are the front shocks.

This is how the transverse spring and right side hairpin radius rod attach to the front axle on the author's track T. As purchased, the front axle already had the mounting plate welded in place. Hairpin radius rods are used, but this same axle bracket could be used for parallel link radius rods. Not seen in the photo is the lower shock mount tab, which is welded to the bottom of the axle and is located in about the center of the photo.

The hairpin radius rod mounts to one location in the frame rail of the author's track T. Since the frame rails taper inward, the manufacturer welds a piece of round tube in place so that the mounting bolt can be perpendicular to the radius rod. If desired, a mounting bracket could be attached to the lower portion of the frame rail rather than mounting through the frame rail.

An easy way to adjust hairpin or four-link brackets is to adjust one end of each bracket to a predetermined point and then pass a bolt through those adjustable ends. Then adjust the opposite ends as necessary and then pass a bolt through them to assure that they are the same length. This is easier than trying to measure the spacing between the holes. Note that the upper and lower adjustments will most likely be different as required for proper wheel alignment, but corresponding bars should be the same length on opposite sides.

### Parallel Link Radius Rods

Because straight axle and dropped axle suspensions look cool and period, but have their drawbacks, hot rod builders find ways to make them better. Parallel link or four-link suspensions eliminate axle twist with a solid-axle suspension. A popular name-brand kit of this type is the 4-Bar suspension manufactured and sold by Pete & Jake's Hot Rod Parts.

Since each end of an axle that is located by split wishbones or hairpins rotates about a central point (where the end of the radius rod is mounted to the chassis), each front wheel experiences a change in caster as the wheel moves up and down. Additionally, since roadways, driveway entrances, and other bumps in the road move each front wheel differently, the caster of each wheel is constantly changing on

its own and in relation to the opposite wheel. These position changes cause the axle to twist. While an I-beam axle will be able to withstand the twisting forces, a tube axle will not and will ultimately break.

As a method of eliminating the twisting axle problem, the 4-Bar was developed. With the parallelogram design, the axle's vertical orientation (caster) does not change as it moves up and down because opposite sides of the parallelogram remain parallel as the system's joints pivot. Held by a parallelogram system on each front wheel, the axle does not experience twisting force. While this is technically not an independent suspension system, it does provide many of the same benefits with use of an I-beam or tube axle.

Using the properties of a parallelogram, a four-link radius rod allows the axle to maintain its caster angle, even though the wheel and tire move up and down.

Parallel link radius rods attach to the front axle by means of an adjuster that threads into each of the parallel bars and is then sandwiched between the ears of a batwing. The batwing is then secured to the axle by the spring perch and lower shock mount. On the back end of the parallel bars is another threaded adjuster. These are bolted to a bracket that is mounted to the bottom side of the frame rails. With two bars on each side of the vehicle and each of them being adjustable in front and back, caster can easily be adjusted to specification.

### Springs and Shocks

As radius rods locate or hold the axle in the correct position, springs support the load between the tires and the chassis and absorb bumps and irregularities in the road. "Shock absorbers," as the U.S. market calls them, actually dampen oscillation, preventing the spring from continuing to expand and contract repeatedly after compressing and releasing once from a bump. What Americans call shock absorbers the British call "dampers," a technically more accurate term. Besides cushioning the ride for the driver and passengers, a secondary task for the springs and shocks is to keep the tires in constant contact with the driving surface. If you've ever watched a tire and wheel roll down a hill, hit a bump and sail upward, the wheels and tires on your car try to do the same thing when they hit bumps. On a car, however, the upward movement of the wheel compresses the spring, which then expands and pushes the tire back onto the road.

### Leaf Springs (Transverse or Parallel)

When a suspension system uses leaf springs, they are mounted transversely (parallel with the axle) or parallel (using two spring packs mounted perpendicular to the axle). As a leaf spring flexes, the distance between the spring eyes will vary. For this reason, at least one end of a leaf spring must be mounted to the frame by a shackle that is allowed to pivot to compensate for the varying length of the leaf spring.

The front suspension components on this T-bucket are relatively easy to see and therefore should be well detailed. The transverse leaf spring has been disassembled, painted, and reassembled. Two features make this vehicle slightly different from many . . . the drum brakes used on the front and Mustang-style steering rather than Vega style. On this version, the drag link (near the center of the photo) moves forward and backward as the steering wheel is turned. This pushes and pulls on the upper steering arm. The lower steering arm mounts the tie rod forward of the axle instead of behind it.

Parallel leaf springs, once common, are not as prevalent as they used to be. While many people may recognize them from the backs of pickup trucks, they were common on the front of GM passenger cars in the 1930s.

Early Ford vehicles and some others used a transverse single spring pack mounted above and parallel to the axle. The spring was secured to the axle with spring shackles secured by spring perches and to the front frame crossmember with U-bolts. As the suspension moved, whether due to bumps or steering input, this design allows the frame to roll in relation to the suspension. As the frame moves from side to side, this causes the drag link to move in relation to the axle, even though the steering wheel may not have been turned by the driver. This process causes the vehicle to oversteer and when combined with greater speed can easily cause the vehicle to spin out. While this general design is still widely used, a Panhard rod is typically used to prevent the frame and body from rolling laterally on the spring, providing more predictable steering. The Panhard rod is a length of steel tubing that has a threaded adjuster installed on each end. One end connects to the axle outboard of the frame rail. The opposite end of the Panhard rod connects to the opposite frame rail. To visualize the concept of what a Panhard rod does, imagine that the axle and front frame crossmember are the two horizontal members of a four-sided box, while the shackles that connect the spring to the axle are the vertical members. Without a Panhard rod, force applied to the top of the box will cause the vertical members to lean one direction or the other and allow the top member to roll in that same direction. A Panhard rod serves as a diagonal brace to prevent this from happening.

Parallel springs, common on early General Motors vehicles, utilize two sets of springs, one on each side of the vehicle. When used on the front suspension, parallel leaf springs typically have a shackle at each end that is attached to the bottom of the frame rails. The axle is typically located on the spring by a locating pin and/or a spring pad. Although the spring with a shackle at each end is similar to the transverse spring mentioned previously, the shackles are situated so that the frame does not roll forward or backward about the axle in the same fashion as the body would roll about a transverse spring. Parallel springs serve to suspend the axle and to locate it, thus eliminating the need for a radius rod. Whether using transverse or parallel leaf springs, it is still necessary to use shock absorbers to dampen and dissipate the energy that is created when a vehicle encounters an uneven roadway.

### Coilover Shocks

Coilover shocks combine a coil spring with a tubular shock absorber to provide a method of load carrying and shock dampening in a compact area. Unlike leaf springs that distribute the load over a larger area, coil springs pinpoint the load. However, in a lightweight vehicle such as a hot rod, this is of little consequence.

To use coil springs (or coilovers) effectively, you must know the required spring rate for the vehicle they are being installed upon. The spring rate is the amount of weight (in pounds per square inch) required to compress the coil

spring 1 inch. While the weights of vehicles will vary for a variety of reasons, including engine and transmission choice, approximate spring rates can be determined by the type of vehicle and past experience of professional chassis builders.

| Coilover Spring Rate (at 1 Inch of Compression) | |
|---|---|
| | **Vehicle Type** |
| 135 pounds | Standard T-bucket |
| 160 pounds | Lightweight pickup |
| 180 pounds | Lightweight fiberglass '28–'32 roadster |
| 200 pounds | Mediumweight '28–'32 roadster, lightweight '28–'32 coupe |
| 220 pounds | '28–'32 coupe, light sedans, '34 roadster |
| 250 pounds | '34 coupe, Mediumweight sedans |
| 300 pounds | Heavyweight '32–'34 sedans |
| 350 pounds | '35–newer roadster |
| 400 pounds | '35–newer coupe |
| 450 pounds | '35–newer sedans |

NOTE: While this information is based upon typical weights of Ford-bodied hot rods, vehicles from different manufacturers of similar year and body style will be similar to those shown. However, the spring rates shown are for comparison only and should not be considered absolute.

The correct spring rate on a pair of coilovers will make a big difference in the ride. To determine the correct spring rate, you would need to approximate the overall weight, which would include the vehicle itself, driver, passengers, and luggage.

Coilover shocks and standard tubular shock absorbers must be mounted correctly to work effectively and to prevent failure. First and most important, the shock must be mounted between the axle and the frame component (frame rail, mounting bracket, etc.) so that the shock travel is perpendicular to the mounting surface. Most shocks used for hot rods have an eye on each end that is slid over a tubular mount or through which a bolt is placed. It is essential that the tubular mount be welded securely in place and perpendicular to shock travel (not necessarily perpendicular to what it is being welded to) to prevent the shock from binding and eventually breaking.

Secondly, the shock absorber or coilover shock should be positioned for optimum efficiency. When mounted in a vertical position, the shock will be at its optimum position. However, that position is not always feasible, or even obtainable in a hot rod. As the shock is tilted away from vertical, it begins to lose efficiency. When mounted at 45 degrees, it has no more than half of its rated capacity. Ideally, shocks or coilovers should be mounted between vertical (zero degrees) and 30 degrees from vertical. At this extreme of 30 degrees, the shock will still be approximately 75 percent efficient.

Lastly, shocks should be mounted so that when at ride height, there is still two-thirds (or more) of the shock's total travel available. This is not only to protect the shock from bottoming out, but from overextending as well.

### Independent

An independent front suspension (IFS) consists of four basic components that are located on each side of the vehicle: spindle, upper control arm, lower control arm, and shock absorber/spring assembly. The spindle is what the wheel rotates about and is held in place by the upper and lower control arms. The outer ends of the control arms include a ball joint that connects to the spindle and allows it to maintain its orientation while the controls arms move up and down as the roadway surface changes. The shock absorber/spring assembly can be a shock absorber with a coil spring, a coilover shock, or a strut-type mechanism.

If you are willing to forego the vintage hot rod look, independent front suspension setups are available for most hot rod chassis. As seen on this coupe, tubular control arms look much better than most of the stock stamped steel control arms.

A variety of independent suspension designs have been grafted to the front frame rails of many a hot rod over the years, with varying success. Whether that success (or lack thereof) is based on aesthetics or improved ride depends on the specific vehicle. While an independent suspension may improve the ride of a particular vehicle, it may simply "just not look right" depending on the components used. Likewise, polished, tubular control arms may look great, but may not improve the ride when compared to a dropped axle

suspension. Although some rodders will disagree, a common belief among many rodders is that IFS does not improve the ride of a lightweight car as much as it does on a heavier car. Without going into the scientific theory and calculations that would be required to prove one way or another, it would make sense that a heavier vehicle would have more to gain from IFS than one that was lighter.

### Stock OEM

There are two basic methods of adding or adapting an OEM independent front suspension to another vehicle. One is to perform a subframe swap and the other is to swap just the front crossmember, control arms, and related hardware. For the front suspension, possible donor cars for independent suspensions form a lengthy list, and they are readily available. Bear in mind that some are more suitable than others. On the positive side, being readily available makes these components affordable. On the down side, what is ugly but hidden by sheet metal on a late model vehicle is still ugly on a hot rod, which often has less sheet metal to cover it. Although they may be functional, stamped metal control arms found on most OEM American-built passenger cars are lacking in the good looks department.

To do a subframe swap, begin by finding a donor car that has a similar track width to the vehicle you are going to be transplanting it into. The donor vehicle's frame is cut at about the firewall, providing IFS, radiator mounting location, and perhaps even an engine and transmission. Then the front frame rails are cut from the recipient vehicle. So far, the process is fairly straightforward and has been done many times. However, the real task begins when the two frame halves are united. Not only must the welding be of very high quality, getting the two sections aligned horizontally and vertically is a task that should not be taken lightly. Before bringing home a subframe, talk to a highly regarded automotive welding shop about the proposed donor and recipient vehicles. Some modern high-performance vehicle frames are of a high-tensile steel that does not respond well to being cut and welded or that must be reattached using a special approach for strength and longevity. Depending on the subframe that is installed, there may not be enough adjustment available to compensate for lack of precision during that installation. If the suspension cannot be aligned properly, it will quickly wear through tires and possibly other components.

Although subframe swaps are not as common on passenger car-based hot rods as they are on vintage (late 1930s through late 1950s) trucks, they are often more involved than the installer anticipates. An incorrect installation may be difficult or impractical to redress.

The second alternative, installing just a different front crossmember and then the additional hardware, seems to be a more prudent solution, especially in hot rods. Maintaining the integrity of the original frame rails makes sheet metal (fenders, hood, etc.) installation much easier, as the original locations are still there. It is also easier to install just a crossmember square between two frame rails than it is to install a much larger subframe. Across the front portion of a pair of frame rails (approximately 2 feet), being off half a degree from square will not be real noticeable. Being off that same amount over the length of a typical subframe could be disastrous. The crossmember should be installed as accurately as possible, but the control arms will typically have enough adjustment to compensate for a less than perfect installation. Unless, of course, you just really screw up . . .

### Aftermarket

Independent front suspension systems available from the automotive aftermarket range from custom-built, one-off pieces using polished and plated tubular control arms to reproductions of relatively common OEM products. Based on the large percentage of them beneath contemporary hot rods, it is safe to say that IFS units based on the '74–'78 Ford Mustang (and its corporate cousins) are the most popular.

Whether you use the stock crossmember from a discarded 'Stang or spend a few dollars for a repop unit, installation is easy, it's strong enough for most hot rods, and, speaking from experience, it works very well. Likewise with the control arms, whether you go the low-buck route and use stock-type stamped steel, step up and use painted A-arms, or go all the way with polished tubular A-arms, the parts are being reproduced. High dollar or low buck, you can still afford new pieces, just in case you don't have a stash of Mustang parts at your disposal.

In most instances, this stock replacement brake caliper and rotor will stop just as well as the more expensive polished and plated calipers and cross-drilled rotors when you are just cruising around town. Now, if you use your hot rod as a road racer, canyon cruiser, or in a gymkhana, more expensive brakes would be in order. Good brakes are important; fancy brakes just look cool.

Unlike Corvair and Pacer brake systems, replacement parts for most GM and Ford vehicles are readily available. It is good to know that you can most likely obtain new brake pads and other related parts, even if you are in an unfamiliar town.

## REAR SUSPENSION

Although the components are different, the principles of locating the rear suspension are essentially the same as those for the front suspension. The rear axle is located by some type of radius rod, weight is supported by springs, and road irregularities are cushioned by shock absorbers. Or, you have an independent rear suspension (IRS), where control arms (lower only) and outer bearing assemblies are used to support the half shafts (axles).

### Rear Axle Housing

Undeniably, the most popular rear axle housing among rodders is the Ford 9-inch, with its sibling 8-inch right behind it in popularity. With the wide variety of vehicles that these rear ends were used in from the factory, there are

After the crossmember is installed for an independent front suspension, you can go low buck or high dollar for A-arms (control arms) and still get the same ride quality. The big question is if you want the parts to be painted or shiny. Since they are bolt-on items, the A-arms can be dressed up or dressed down.

Available in a variety of widths in stock form and in virtually any width as a custom unit, the Ford 9-inch is probably the most common rear end in the hot rod world, if the not the automotive world. Find one the correct width for your application (or have one cut to your specs), add your choice of gears, brakes, and brackets, and you are on your way toward being able to put some serious horsepower to the ground.

many that are the correct width for many a hot rod without modification. Additionally, there are many stock widths that can be narrowed as required. Ten- and 12-bolt rear ends from General Motors are the next most commonly found between the rear wheels of a hot rod, with the majority of these being in GM vehicles. Ford rear ends, however, are commonly found beneath the back end of just about any make or model.

### Wishbone Radius Rods

Like in the front of early hot rods, wishbones are commonly found in the back as well, since they were stock equipment that has merely been modified. Rather than bolting to the axle, rear wishbones are typically welded to the axle housing. Removing the stock pivot arrangement involves cutting off the front ends of the rear 'bones and then welding a threaded bung in place. An adjuster is then added to allow for slight adjustment of the rear axle (pinion angle, wheelbase, and squaring the axle with the frame). The front of the wishbones can be mounted to the frame rails or to a crossmember just slightly outboard of the driveshaft. This latter arrangement is more desirable when the vehicle will be used for drag racing or lakes racing. Where it gains in traction, it loses slightly in passenger comfort.

### Hairpin Radius Rods

Rear hairpin radius rods are very similar to the version made for the front of the vehicle. The only significant difference is that the rears are typically somewhat longer than those used for the front. They mount to a mounting plate that is welded to the rear axle housing.

Much like on a contemporary Ford rear axle housing, the brake backing plates are bolted to the axle flange. In the stock configuration, the rear wishbones were secured to the axle with the same set of bolts.

For this contemporary application of split wishbones being mounted to a Ford 9-inch rear axle housing, the stock mounting holes are used to secure a custom-made bracket. This bracket will then be welded to the rear axle housing.

This rear wishbone has had the center pivot point removed (split) and now requires a method of attaching the front end of the split wishbones to a forward position on the chassis. A threaded bung will be welded in place and then an adjustable rod end threaded into it. The jam nut will be tightened against the threaded insert to maintain the position of the rod end.

The front end of this rear wishbone will be mounted slightly inboard of the frame rails to a round tube crossmember. To do that, two tabs that are notched to fit the crossmember are bolted to the rod end.

Use a mallet to tap the threaded bung into the end of the wishbone, making sure that it is flush. Use some sandpaper to remove any rust from the wishbone and/or the threaded bung.

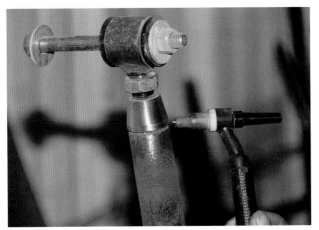

Then weld the bung in place with your favorite piece of welding equipment. A TIG welder will look the best, a MIG will probably be the easiest for the beginner, but a gas welder has been used to do this on countless hot rods. Just be sure that a good weld is achieved, no matter what kind of welder is used.

The threaded bung is tack welded in place at first, but it will eventually be welded all the way around.

### Ladder Bars

Designed for racing, ladder bars help to reduce rear axle twist during acceleration. Most ladder bar setups are attached to the rear axle housing near each outboard end of the housing and then to the transmission crossmember just outboard of the transmission output shaft, much like rear wishbone setups. The big difference between rear wishbones and ladder bars is that each ladder bar is made up of two tubes. The two tubes are approximately 4 to 6 inches apart where they connect to the rear axle housing, but taper so that they use one threaded adjuster or bushing at the front mounting point. The ladder bar pivots from this front mounting point, allowing the axle housing to travel up and down as required. Adjusters on the two tubes near the axle mounting points allow the suspension to be adjusted slightly, mainly to control wheel hop and axle twist during acceleration. Since ladder bars are associated with racing, they typically have strengthening gussets between the tubes.

### Four-Link Radius Rods

Rear four-link radius rods can be one of two types: parallel or triangulated. Parallel link radius rods locate and control rear axle movement just like those designed for the front axle, using the parallelogram principles. When using this with coil spring, transverse leaf springs, or coilovers, it is necessary to use a rear Panhard rod to prevent the body from rolling in relation to the rear axle. Triangulated four-link setups are similar to parallel setups for the lower bars. They mount to an axle bracket at the rear and to the frame in front. On parallel setups, the upper bar is, as you would guess, parallel to the lower, and it mounts in similar fashion, albeit above the lower bar. On a triangulated setup, the upper bars mount to the rear axle housing slightly outboard of the center section and then flare outward to mount near the frame rails at their forward location. Due to the triangulation, this setup eliminated the need for a Panhard rod.

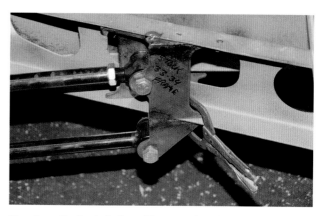

Depending on the chassis, the forward frame mounting points for a rear parallel four-link radius rod will be the frame rail itself, a bracket mounted below the frame rail, or a bracket mounted to the stock center crossmember. The one shown is for a stock crossmember in a '33–'34 Ford.

The axle mount for a rear parallel four-link will fit around the bottom of the axle tube; it usually includes lower coilover shock mounts on the back side. When positioning these brackets on the rear axle, make sure that they are positioned so that the bars pivot freely and are perpendicular to their mounting bolts. In other words, mount the frame mount first, and then mount the rear bracket where it needs to be. Remember that if coilover shocks or a transverse leaf spring are used with a four-link, a Panhard rod must be used to minimize lateral movement of the body in relation to the chassis.

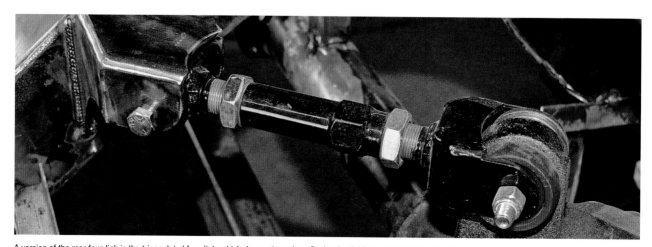

A version of the rear four-link is the triangulated four-link, which does not require a Panhard rod. Like a parallel four-link, the lower bars mount to the rear axle and the frame rails. The upper bar, however, mounts near the middle of the rear axle housing and diagonally to a point on the frame rails.

## Leaf Springs (Transverse or Parallel)

When used on the rear end of a vehicle, transverse leaf springs typically have a much higher arch than springs designed for the front of the vehicle. Most everything else that applies to front transverse leaf springs applies to the rear as well.

In similar fashion, parallel leaf springs are very similar in characteristics to their forward counterparts. Perhaps the

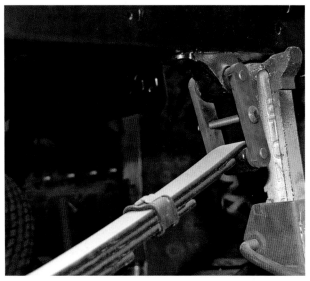

At the back of the spring, a shackle connects it to the frame rails. With the shackle being at a fixed location on the frame rails and allowed to pivot, the leaf spring is allowed to flex.

Transverse leaf springs are still fairly common beneath the rear end of many hot rods, although not all look as nice as this polished piece. The somewhat strange looking device is a form of lever shock absorber. The cylindrical discs mount to the frame rails via a standoff bracket. The opposite end connects to a Heim joint that ultimately connects to the rear axle.

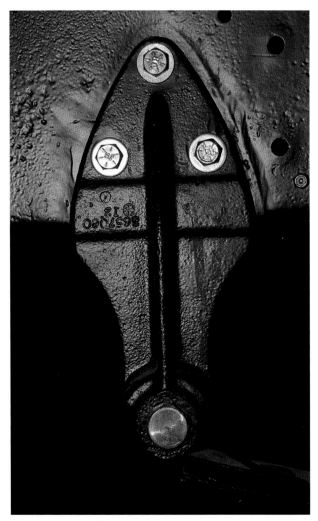

Most commonly found beneath pickup trucks, both old and new, parallel leaf springs provide a decent ride and can support a significant load. As with any spring, the allowable load and ride quality will be dependent upon the characteristics of the spring being used. To lower the rear of this Advance Design Chevy truck, the rear axle has been repositioned atop the springs.

In the front of that same leaf spring, the spring eye fits over a fixed mounting point.

The rear axle is secured to the parallel leaf springs by a pair of U-bolts and a spring pad on each end. By placing the axle above the springs (rather than in the stock location below the springs), the vehicle is lowered by the diameter of the axle housing.

only difference is that there is usually only a shackle at the rear of the spring pack when used on a hot rod. The front of the spring is usually mounted by a bolt through a spring eye in combination with a bracket mounted to the frame that sandwiches the spring. A shackle between the spring and the rear mounting point allows for the flex of the spring.

## Independent

Although designs will vary slightly from one manufacturer to another, most IRS systems share basic component concepts. There is a housing (contains the ring and pinion gears); half shafts (the axles that transmit power to the rear wheels); bearing supports (locate the outboard end of the half shafts); and control arms (connect the bearing supports to the housing). A crossmember to which the housing is attached allows the entire assembly to be installed between the vehicle's rear frame rails. Coilover shocks are typically the spring of choice and are mounted between the lower control arms and a bracket located on the crossmember.

An independent rear suspension will most likely improve handling, cornering, and traction in any vehicle. Since OEM units are normally found in vehicles with more than ample horsepower, breakage of an IRS is typically not a problem.

### Stock OEM

Undoubtedly, there are other IRS units available from salvage yards, but Jaguar and Corvette are the two most commonly thought of for use in hot rods. Jaguar IRS units typically utilize coilovers, while the Corvette utilizes a transverse leaf spring. Benefits of aluminum Corvette units are that they are more plentiful; bearings, seals, and other parts are more easily

obtainable; and they are lighter than the steel Jaguar units. A downside of the Corvette unit is that it has a relatively wide wheelbase that simply will not fit beneath hot rods from the early 1930s; however, they do fit nicely under fat fendered hot rods.

### Aftermarket

Several manufacturers offer their versions of IRS units, utilizing the basic concepts and components mentioned previously. A benefit of these aftermarket units is that they can often be built in custom widths to fit beneath an early hot rod or one that uses very wide rear tires to get that Pro Street look.

### Air Bag

Air bag suspensions use the same basic principles of suspension design to locate the axle or A-arms as their nonair counterparts. The biggest difference is that inflatable "bags" are used in place of springs and shocks. With most air-suspended vehicles utilizing on-board air compressors and regulators, air pressure can be monitored and adjusted at the push of a button or flip of a switch. This allows the driver to lower the vehicle literally to the ground when parked, raise it by adding air to "cruising altitude," or adjust ride height as necessary based on payload.

Although the air bag suspended components are similar to their more conventional counterparts, they are not always directly compatible, due partially to the differences in sizes of fully inflated and deflated air bags. For some vehicles, retrofit kits are available for installing air bags on conventional suspension systems.

# PROJECT 1
# Installing an Aftermarket Transmission Crossmember

Any hot rodder who drives significant distances to rod runs or on vacation knows that an overdrive transmission makes long miles easier. With the seemingly everyday increase in gasoline prices, having that same overdrive transmission is even making sense in rods that are limited to around-town use. The down side of an overdrive transmission is that it is typically bulky in comparison to its nonoverdrive counterparts. When you combine this with the relatively confined area of an early Ford chassis, many rodders simply continue running their nonoverdrive transmissions rather than make what is now an easy upgrade.

Since the transmission crossmember is what provides most of the rigidity in an early Ford frame, you can't just arbitrarily start cutting away metal to make room for a larger transmission. Additionally, if you are retrofitting an overdrive tranny into an existing rod, you really don't want to be required to remove the body from the frame or have to do extensive paint touch up.

Whether you are installing the popular GM 700R4, a Ford AOD tranny, or one of the other large-bodied transmissions, Morfab Customs has a practical solution. Keith Moritz and his crew have been manufacturing a bolt-in kit for installing these larger transmissions into earlier

fat-Fords for several years, but they have recently added the kit for the '41–'48 Fords. Like the previous kits, it can be installed with or without the body in place. An added bonus of this particular application is that the kit includes the master cylinder mount, which is drilled for both standard master cylinders or for those utilizing a power booster. For the most part, this kit is a true bolt-in, requiring no welding. Instructions included with the kit show you where to cut away the original crossmember, grind off a few rivet heads, and then bolt the replacement pieces in place. It just doesn't get much easier than that. The kit is slightly different, depending on which transmission you are using, so be sure to specify which transmission you have when you order. This kit cannot be used with stock wishbones.

This kit can be installed with the body in place, but for clarity is being done with the body off the chassis. For these photos, the chassis is starting in the right side up position. It will be turned over a couple of times during the procedure for simplicity, but we'll let you know when. Follow along as Keith Moritz installs a Morfab Customs overdrive transmission installation kit (it also allows for installation of nonoverdrive transmissions, such as the GM 350TH) into a '41–'48 Ford chassis.

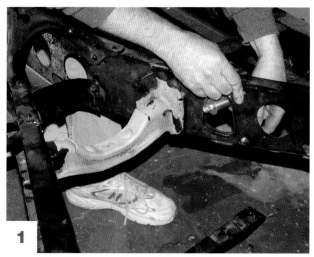

**1** The first step is to remove the brake master cylinder and pedal. If this is being done on a drivable vehicle, be sure to drain the brake fluid before this step.

**2** Body mount braces that are attached to the outer frame rails and the X-member need to be removed temporarily. They will be reused, so don't trash them. They are riveted in place from the factory, so grinding the heads off and then knocking them out is the typical procedure. On this particular frame, the body mounts have already been removed and reattached with bolts and nuts. Regardless of how they are attached, they need to be removed.

**3**

The stock transmission mount needs to be removed. Keith begins this process by using a small grinder to grind off the rivet heads. With the heads ground off, a hammer and punch can be used to drive the rivets out of their original location.

**4**

Theoretically, with the rivets removed, the factory transmission mount would fall right out. However, this frame is a minimum of 60 years old, so it may be a bit cantankerous. If this is the case, some persuasion from a Ford hammer (a.k.a. BFH) may be appropriate.

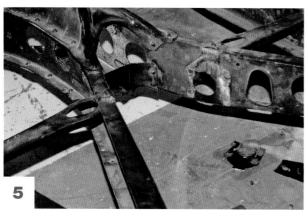

**5**

This is essentially what the process would look like from the topside with the stock transmission mount removed.

**6**

Not that it has to be, but the chassis has now been inverted to make removing a portion of the transmission mounting plate easier. Measure back 1 ¼ inches from the second pair of rivets from the front of the bottom of the factory mounting plate and scribe a line across the plate.

**7**

Cut the transmission mounting plate along the inside of both legs of the X-member. A plasma arc cutter makes easy work of this, but it could be done with a die grinder, reciprocating saw, or cutting torch, depending on what you have available. Keith uses a die grinder to make the lateral cut across the transmission mounting plate. Just be sure that you don't cut through the frame.

**8**

With the factory transmission mounting plate out of the way, grind the heads from the two forward rivets that secure it to each of the legs of the X-member.

**9** Punch out the rivets and then use a hammer, chisel, and pry bar to remove these leftover pieces of the bottom factory transmission plate from the legs of the X-member.

**10** Patina indicates the portion of the transmission mounting plate that has been removed. A new bolt-in plate will mount to these original rivet holes and therefore will be removable for future tranny service. The X-member replacement panels bolt to existing rivet holes before the sides are cut out of the factory X-member. After grinding off the heads from the appropriate rivets, Keith drives the rivets out with a hammer and a punch.

**11** In some cases, the rivets may need to be drilled out. Just make sure that you drill the rivet and none of the chassis.

**12** Now after mounting each of the X-member replacement panels with bolts provided in the kit, mark a line on the original X-member where it needs to be cut away. Use a plasma arc cutter or die grinder to cut the original X-member.

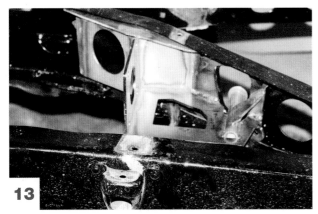

**13** This is the driver's side X-member replacement panel, complete with the master cylinder mount. Note that the frame is still upside down in this photo. The bare spot on the flange of the frame rail is where the previously removed body mount was located.

**14** You can go ahead and reinstall the driver's side body mount at this time.

**15**

The passenger side X-member replacement is installed using the same basic procedure. Begin by grinding off the appropriate rivet heads.

**16**

Drive out the rivets with a hammer and punch.

**17**

Mount the X-member replacement panels and secure with the bolts provided in the kit. Mark the original X-member as necessary and cut with a plasma arc cutter or die grinder. Remove the original pieces and grind the frame smooth as desired.

**18**

These replacement X-member panels provide much needed room in the transmission mounting area. Since this particular vehicle is a new construction, the entire frame will be primed and painted later. If you are using this kit to retrofit a complete vehicle, the new pieces should be primed and painted before installation.

**19**

The transmission mounting plate is what the transmission actually bolts to and is removable for transmission removal. Since the mounting hole pattern is different for Chevrolet and Ford transmissions, be sure to specify which you plan to install when you order the kit. The transmission mounting plate is installed the same way in either case.

**20**

On the passenger side, the mounting position for the body mounting bracket is no longer there. With the body mounting bracket secured to the outer frame, drill a new hole in the bracket where it aligns with the replacement panel that is part of this kit, and secure it with a bolt and nut. The body mount must then be trimmed where it encroaches into the new space created by installation of this kit.

**21** Keith first uses a die grinder to trim the body mount to the appropriate length and then bevels the end slightly.

**22** This is what the installation looks like before the installation of the transmission. Very straightforward and easy to accomplish.

**23** Even though the body may be considered fat, the skeleton of a 1940s Ford is anything but spacious. The 700R4 is the bulkiest of the contemporary automatic transmissions being used by street rodders, so if it fits, most everything else will too. With different bottom plates, most any passenger car transmission can be installed with this kit. Note that the brake pedal shown is not actually the one to be used with this transmission.

# PROJECT 2
# Building a New Chassis

**W**hether you are resurrecting a vintage chassis, starting anew with a set of reproduction frame rails, or plan to purchase a perimeter frame or complete chassis from a chassis builder, a look at how the professionals do it will be helpful. To see how a pair of bare frame rails become a perimeter frame, I followed along as Keith Moritz at Morfab Customs builds a '33–'34 Ford frame with an IFS crossmember.

**1**

Morfab Customs uses ASC reproduction frame rails for the chassis and perimeter frames the company builds. This is a pair of frame rails for a '33–'34 Ford just as they come off the delivery truck . . . a wooden block between them and bound together at the front and back.

**2**

The frame rails are placed into the chassis jig, aligned, and then tack welded in place. All welding of the chassis will be done with the frame rails in the chassis jig to ensure that the chassis is dimensionally square and has no twists.

**3**

Due to the shape of these particular frame rails, the boxing plates are made of three separate pieces for each side. Keith has already installed the middle and rear boxing plates and is prepping the front passenger boxing plate for installation.

The boxing plate is checked for appropriate fit. Note that the forward portion (approximately 12 inches) of the frame rails will be left unboxed on this frame, which will be used on a fendered car. On fenderless hot rods, this same area is often boxed also, but that is the builder's choice.

Keith uses a locking clamp near the middle to hold the boxing plate in place prior to it being tack welded in place.

Keith finesses the placement of the boxing plate slightly to make sure that it fits properly . . .

. . . and then adds another locking clamp to hold the front portion of the boxing plate in place.

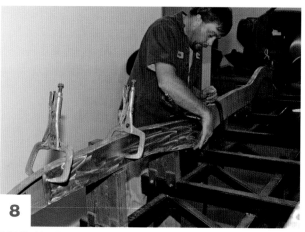

**8**

A third locking clamp is added to hold the rear portion of the boxing plate in place. Taking the time to ensure the boxing plates are aligned prior to welding will eliminate many problems later.

**9**

With the boxing plate accurately clamped into position, it should be tack welded in place in several locations.

**10**

Rather than lengthy continuous welds, short tack welds spaced 12 to 18 inches apart are best for tacking the boxing plate in place.

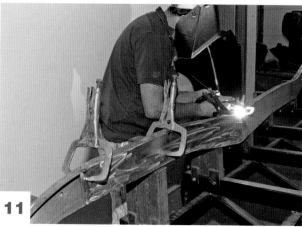

**11**

Notice the rigidity of the chassis jig. To be able to produce dimensionally square chassis with consistent results, a chassis jig is a must. A square original frame, lots of steel tubing, and countless hours spent welding are required to build a jig.

**12**

Having a chassis jig that is also on a rotisserie makes working more comfortable and therefore yields better results. Having visual access as well as physical access to what you are welding is a great advantage in doing the job well.

**13**

Keith has already welded the rear two boxing plates. Prior to the chassis being finished, the boxing plates will be welded around their complete perimeter, except where they span the open portion of the frame rails at the front and back.

**14** Using a combination square, Keith marks a line on the inside of the front portion of the frame rails to indicate the correct location of the center of the new front crossmember. Verify locations with the instructions provided with the components you are using.

**15** The crossmember centerline location is transferred upward from the appropriate stock hole and just over the upper corner of the boxing plate.

**16** After the process is repeated on the passenger side, the individual lines are transferred across the top of the frame rails by tracing along a metal straightedge. By spanning across both rails, it is easier to notice if the proposed centerline is not square with the frame rails.

**17** However, to double-check and verify that the crossmember will be installed squarely, clamp the metal straightedge in place . . .

**18** . . . and then check measurements to a specific location on the rear of the frame rails.

**19** As with most all chassis construction, measurements are checked on both sides of the vehicle. Yes, they should be the same on both sides.

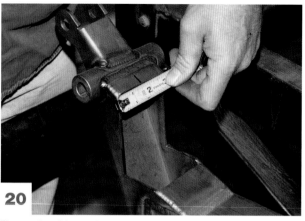

**20**

The centerline of the actual crossmember is now determined and then a reference mark placed on the upright portion that will be welded to the outside of the frame rails. This mark will align with the marks on the top of the frame rails.

**21**

Keith uses a piece of scrap steel tubing and a bar clamp to secure the crossmember until it is tack welded in place. With the clamp near the middle of the crossmember, it is easier for one person to verify that the crossmember is properly aligned with the frame rails.

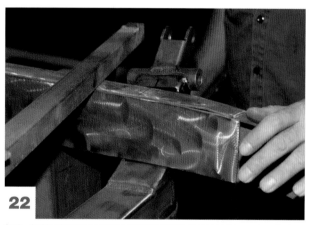

**22**

Again, double-check that alignment marks are indeed aligned. It is much easier to make corrections now, rather than after the crossmember is welded in place.

**23**

By the light of the welding torch, the alignment marks are a little easier to see on both the crossmember and the frame rail. The bar clamp and the steel bar adjacent to it is a clamping mechanism used to squeeze the rails inward slightly so that the crossmember fits accurately around the frame rails. Several tack welds are done on each side of the crossmember to make sure that it does not move.

**24**

This is what the chassis looks like with the Heidt's IFS front crossmember tack welded in place.

**25**

The boxing plates and front crossmember have now been fully welded in place. For a quick and dirty chassis, the welds could be left as is. However, all chassis to leave Morfab Customs, as well as most if not all other reputable chassis shops, will have the welds ground smooth.

**26**

The tubular center crossmember has already been welded together and now awaits installation between the frame rails.

**27**

If the center crossmember tubes are to be bent, the tubes must start out longer than the distance between the frame rails. After the crossmember tubes are bent and assembled, the main tubes are cut to the approximate length to fit between the frame rails. They will be trimmed to fit precisely later in the process.

**28**

For a '33–'34 Ford chassis, Morfab Customs uses three boxing plates on the inside of each frame rail, which results in two seams per side. When the welds are ground smooth later, the seams will essentially disappear.

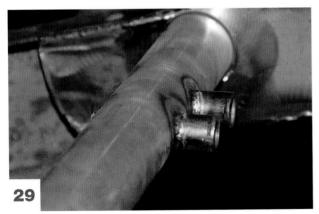

**29**

This chassis will use coilover shocks in the rear, so the rear crossmember is designed to serve as the upper mount. Each side of the rear crossmember has two mounting tubes to mount the upper end of the coilover. Having two locations provides a certain amount of adjustability of ride height and dampening.

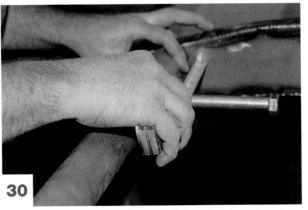

**30**

The rear crossmember in this case is a round tube that extends to slightly inside of the holes in the boxing plate. While this allows for a full circumferential weld and a stronger rear crossmember, lateral location must be checked and verified before the crossmember is welded in place. The distance to the driver-side boxing plate from one of the coilover mounting holes is measured.

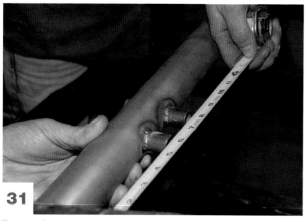

**31**

The same distance is measured on the passenger side of the chassis. If the measurements differ, the lateral location is adjusted until the measurements are the same on both sides.

**32**

With the rear crossmember centered side to side, it must now be rotated so that it is level with the frame. When building a chassis jig, design it so that the chassis is level, regardless of the profile of the frame rails or the rotation of the rotisserie, if so equipped. By using an angle gauge on a bolt extending from the upper shock mounting hole, Keith finds that the rear crossmember is correctly oriented.

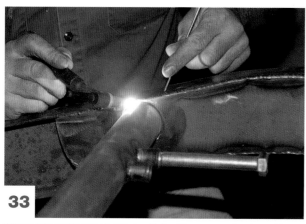

**33**

With the lateral and angular locations verified and the crossmember clamped in position, the rear crossmember can now be tack welded in place so that it will not move laterally or rotate.

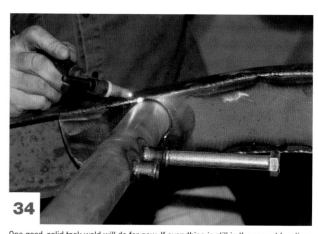

**34**

One good, solid tack weld will do for now. If everything is still in the correct location, more welding will be required, but if anything moved, only one tack will need to be cut loose.

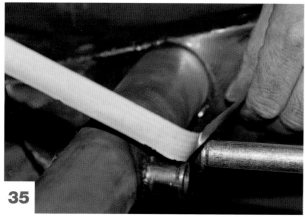

**35**

A quick measurement from side to side will verify that the crossmember is still in the correct position. The angle of the rear crossmember should be checked too, just to make sure that the coilovers do not end up being put in a bind.

**36**

With the location double-checked, Keith places a few tack welds on the passenger side of the rear crossmember and a couple more on the driver's side as well.

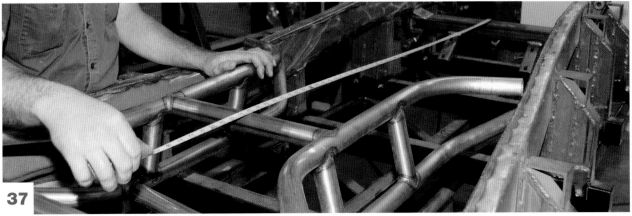

**37**

Now it is time to locate and position the center crossmember. With the crossmember in the approximate position, Keith takes a measurement from the front crossmember to the upper tube that connects both sides of the assembly. This measurement may differ between manufacturers but will be consistent with any one manufacturer's chassis. When the center crossmember is located at the correct distance from the front, the tubes should require only minor grinding to fit properly.

**38**

Keith uses a handheld grinder to trim a slight bit of excess from one tube of the center crossmember.

**39**

Besides not trimming off too much metal, it is important to keep the ends of the tube ground to the correct angle in relation to the frame rails and the boxing plates. A little bit of gap is acceptable, as the tubes will be fully welded to the boxing plate.

**40**

Dave Hidritch, owner of this underconstruction chassis, stopped by the shop to check progress and lend a hand if necessary. As with most hot rod building chores, an extra set of hands comes in handy on occasion. The fit is close, but not quite perfect yet.

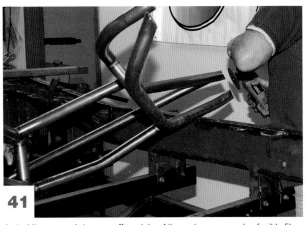

**41**

Just a bit more needs to come off one tube of the center crossmember for it to fit correctly. A quick hit with the grinder and that is done.

**42**

Now that the crossmember is trimmed properly, it will pretty much stay in place by itself, but it needs to be clamped prior to welding. Keith uses a piece of square steel tubing that spans the frame rails and a couple of bar clamps to hold it in place.

**43**

Before tack welding the center crossmember in place, Keith takes some measurements once again, just to make sure everything on this chassis is as it should be.

**44**

Since this chassis is being built mostly after-hours to work with my available photography schedule, Keith had not yet ground down the welds on the boxing plates. The grinding could be done before or after trimming the center crossmember to fit.

**45**

Welding the boxing plates and then grinding the welds smooth are probably the two most time-consuming, and therefore the most expensive, portions of the chassis build.

**46**

To remove the excess weld, Keith uses a 36-grit sanding disc on an angle head grinder. When that initial grinding is completed, he switches to a 50-grit disc to finesse the finish a bit. In both cases, keep the grinder head moving to keep any heat buildup to a minimum.

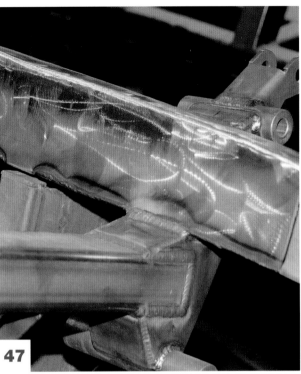

**47**

With the grinding completed, the chassis will look something like this. The grinding will provide a suitable surface for an application of epoxy primer after the chassis is cut from the jig.

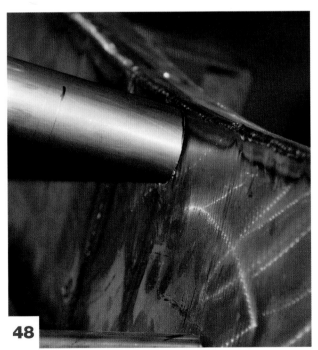

**48**

Regardless of when the rest of the boxing plate welds are ground smooth, the area in the immediate vicinity of the center crossmember must be ground before the crossmember is welded in place. Doing this will allow for better access to grinding the welds and for providing a better weld of the center crossmember tubes to the boxing plates.

**49**

The center crossmember can now be tack welded in place, all welding completed, and welds ground smooth. The chassis would then be cut out of the chassis jig, cleaned with wax and grease remover, and then protected from the formation of surface rust by an application of epoxy primer.

## PROJECT 3
# Safety Wiring Brake Rotors

**V**irtually all sanctioning bodies that govern automotive competitions require that rotating assemblies that are bolted together utilize safety tie wires. The reasoning for safety wires is to prevent the assembly bolts from backing out and allowing the rotating assembly to part company with the rest of the vehicle. While you may never actively compete with your hot rod, safety wiring the (disc) brake rotors to the hub is still a good idea. Grant Scott at Morfab Customs explained the relatively easy process.

**1** The heads of the bolts used to secure the rotor to the hub must be drilled for safety wire. Although you could drill the bolt heads yourself, it would be more cost-effective to purchase bolts that are already drilled. A length of safety wire is threaded through the bolt head.

**2** Both ends of the safety wire should be pulled in the same direction toward the adjacent bolt. The safety wire should be positioned so that the portion that passes on the outside of the bolt head will prevent the bolt from turning in the loosening direction. In other words, if the bolt is a right-hand thread, the portion of safety wire that runs along the outside of the bolt head should be on the right side of the bolt.

**3**

Although you can use ordinary pliers to twist the wire, special safety wire pliers allow you to pull the wires taut and turn a handle to twist the wire in a uniform fashion.

**4**

As the twisted wire approaches the next bolt head, quit twisting the wire. Pass one end of the wire through the hole in the bolt head, while the other end of the wire passes on the appropriate side of the bolt on the outside. Bring the two ends of wire together again, and begin twisting the wires as before.

**5**

Continue this process around the assembly and through each of the bolt heads until you reach the last bolt. Pass one end of the wire around the bolt and the other end through the bolt head, and then twist them together for approximately 1 inch. Cut off the excess wire just inside the end of the twisted-together part. Using a needle-nosed pliers, turn the end of the twisted-together wire back on itself in order to hide the sharp end so it doesn't poke somebody and get blood all over everything.

# Chapter 3
# Hot Rod Body Repair

There is no denying that building a hot rod is about personalization and making a statement. Even if your perfect hot rod is subtle and subdued, it is still a personal statement. Whether you are resurrecting a vintage body that has spent part of its life in a creek bed or are hot rodding a 100-point restoration, there is still work involved to make it *your* hot rod. Either extreme or somewhere in between, you will most likely have the opportunity to do some repair work and some modifications to your hot rod body. In some instances, you will need to repair before you can modify, while others can be addressed simultaneously.

Perhaps the best advice anyone can give here is to do a thorough inspection of the body and related components before performing any work on them. We all know that when your project vehicle was sitting in someone else's garage and

had their name on the title, it looked pretty pristine. It sure is funny how ownership can change your opinion . . . You should verify the *actual* condition of the pieces and parts to see if they can be used as is, used with repair, should be replaced, are swap meet material, or are dumpster filler. If the pieces are in decent shape, needing little to no repair, that is a good thing. If they need repair, do you have the resources (time, talent, equipment) to make the necessary repairs? Are replacements available if the parts require work that is beyond your means? If you are going to replace the part, is it worth anything to anyone else who may be willing to pay you cash or trade some parts or labor for it? Determining realistic answers to these questions will eventually save you extra work, minimize a growing inventory in your garage, and potentially save you some money.

No matter how much you may want to build a hot rod project, many first-time builders start with a project that offers a low buy-in, only to find that it requires more experience and/or money than they can put into it. As you walk around a potential project vehicle, you should be asking yourself if you have the ability to repair what needs to be repaired and whether parts are available for the items that are beyond repair.

It looks like a simple flat piece of sheet metal has been used to "replace" the area between the deck lid and the rear subrail. Not bad, but a reproduction panel would be a better choice. The rear quarter panel needs a sizable patch panel. The rear fender has a common tear, gobs of body filler, and appears to be twisted. You can probably buy a new fender for less than the expense of repairing this one.

Even though the running board can be pitched, the running board splash apron does not look to be in bad shape. Is the paint bubbling up? If it is, there is serious rust beneath it.

OK, there was a gaping hole in the back of the quarter panel, and now we see there is rust out in the front portion of the same panel. Many a body man has had less to start with when building a hot rod, but do you want to go to that much trouble? Since the character line still exists in the front and back of this area, you just need to figure out how to bridge the gap.

The hood side doesn't look too bad, although it is not as flat as it should be. The louvers all seem to have been tweaked slightly and irregularly . . . a point that would require dedication to straighten completely. Looks to be lots of body filler in the fender, all of which should be ground out before having the fender chemically stripped, should you choose that route.

Gee, it doesn't look too bad from the front. Now you are probably envisioning it sitting 4 to 6 inches lower, with some shiny "Arrest Me Red" paint. A small-block with three deuces under the hood, oh man, this is gonna be so neat. . . . Now is when your hot rodding buddy needs to smack you and bring you back to reality.

Replacement cowl panels are available for this make and model, so if this was the only body repair to be made, it would be easy to resurrect with some time and a MIG welder.

The stock sun visor is probably the best remaining panel on this potential project. But, a sun visor is a heck of a place to start a hot rod project.

I guess that is one way to patch a panel, but certainly not the desired method. Rather than simply tack welding a piece of scrap metal over the hole, cut the replacement panel and the hole to the same size, and then butt weld the two together. You may still require a skim coat of filler, but not as much as this "repair" will take.

you ask? If a potential hot rod project has little or no paint on it, it will probably have some amount of rust on it. But you can do a little bit of investigating by doing some discreet scraping with a pocketknife to gain a better idea if the metal underneath the patina is reasonably solid or if it is paper thin. On the other hand, if the vehicle is sporting a new coat of paint, it may be full of body putty. Body filler itself is not all bad, but there are right and wrong ways to use it and you need to know what's there. When it is covered by a coat of paint, it is very difficult to tell what lies beneath. Of course, if you are familiar with the freshly painted vehicle and trust the seller or it is a fiberglass body, then rust is not a problem.

While this rat rod includes a uniform coating of patina, it would be easy to distinguish between mere surface rust and serious rust through. However, what lies beneath the quick coat of primer where a piece of metal was welded in is anybody's guess.

That thorough inspection can be disheartening at first, but it will ultimately save you some grief. True, you may find that what you thought were perfect fenders are actually full of body filler that is beginning to deteriorate. However, you must admit that it is better to find that out now, early in the project, rather than later when think you are all ready to start spraying paint.

## RUST/PAINT REMOVAL

Rust is one of those items that must be removed completely whenever you are building a hot rod. It may move and grow slower in some areas of a vehicle than others, but if it is there and you do not remove it, it most likely will come back to haunt you. For this reason, many rodders prefer to begin hot rod projects with vehicles that have readily apparent surface rust, rather than one with a shiny coat of new paint. Why,

Rust can be removed by using chemical methods or mechanical methods. Both methods have pros and cons, with the amount of rust to be removed often being the critical factor when choosing which to use. If neither of these methods is suitable, it may be more practical to simply cut out the old and rusty metal and then replace with new.

### Chemical Stripping

For removing large areas of rust (whether it is relatively minor surface rust or more substantial) and/or paint, chemical stripping has the least amount of potential side effects. If you have a steel body, doors, fenders, and such that can be completely disassembled, having them chemically dipped is the most practical method of getting rid of rust. For this method to be the most effective, disassembling the components to be dipped to their most basic component is the key. If the doors are not removed from the body shell, the hinges cannot be adequately cleansed. If trim, emblems, or other items are not removed, the sheet metal behind them will not be stripped completely. Any wooden supports, glass, rubber, or anything that is not steel should be removed from the parts being dipped, as it will need to be replaced anyway if it gets dipped. Additionally, if you are aware of any thick layers of plastic body filler, you should try to remove the filler prior to having the part dipped. The body filler will be easier to remove with a grinder when it is solid, but it will be soft and difficult to remove after being dipped—trust me on this . . .

These are two products that can be used to remove paint from automotive bodies. The product on the left is designed to remove paint (a collection of chemicals) from fiberglass (another collection of chemicals), without damaging the fiberglass. The product on the right is designed to removed paint and primer from metal surfaces. These should be used for stripping paint off single panels, rather than an entire vehicle.

While this collection of parts could use a bath in the chemical stripper's tank, what is there looks pretty solid, with very little if any rust through. Most of the sheet metal and a rolling chassis make this a good place to start a hot rod project. The buy-in may seem high, but this project would be easier to complete than other projects for less money.

The liquid chemicals in a stripping tank get everywhere that is not sealed when a body or other parts go in. Rust will be removed from the inside, the outside, as well as all of the little nooks and crannies. When the dipping process is completed, what comes out of the tank will be bare metal. If the metal was dented, it will still be dented. If the metal was rusted through, there will now be a hole. If the rust was only on the surface, the metal will now just be a little thinner. However, the good thing is that you will know exactly what needs to be straightened, what would have rusted away eventually, and what is still solid.

When automotive bodies and parts are dipped, they are first submersed in a "hot tank" that is filled with a caustic solution. This is to remove paint, wax, grease, and other chemicals from the metal and will take between four and eight hours, depending on the amount of buildup on the metal. The metal is then removed from the hot tank and rinsed with plain water until all of the caustic solution is removed, typically between three and four hours. Next, the parts go into a different tank that is filled with de-rusting solution. While in this tank, the parts are connected to an electrical charge. The current is charged so that it pulls rust particles away from the parts, rather than pulling the particles to the parts as happens in chrome plating or powder coating. Depending on the condition of the metal and the amount of rust, this step may take 20 to 40 hours.

When the remaining metal is removed from the de-rusting tank, the parts are thoroughly rinsed again with plain water to neutralize any remaining de-rusting solution. Parts that are chemically stripped should be primed with epoxy primer as soon as practical to prevent the formation of surface rust. A benefit or drawback of chemical dipping is that the stripping solution will get to all surfaces of the parts that are immersed. This will remove all rust, but it may leave areas of good metal with no protection if you cannot access them to apply epoxy primer and paint or undercoating.

The door on this past project has been dipped to have paint and primer removed. If the panel is solid, it will look like new metal when it comes back from the stripper. On the other hand, if rust is present, it will be removed.

## Mechanical Stripping

Other methods of rust and paint removal are labor intensive, but are more readily available in some areas than the aforementioned chemical dipping process. Two of the most common processes are media blasting (for large areas) and grinding or sanding, which should be reserved for smaller areas.

## Media Blasting

A great way to remove paint, peeling chrome, and rust is by media blasting. In the early days, this process was known as sandblasting, but there are now significantly better materials to use, hence the more correct name of media blasting. Sand can quickly put heavy scratches into light material and will produce significant heat, causing sheetmetal panels to warp. Silica sand is much finer than river sand; however, it has drawbacks as well. Using any kind of sand for blasting purposes will create an extremely fine dust, which is known to cause silicosis from inhalation. This extremely nasty side effect can be avoided by using any of a number of alternative products.

Four very important things to remember when media blasting are (1) mask the area that should not be blasted, (2) use the appropriate blasting media, (3) remove all of the blasting media when the job is complete, and (4) wear proper safety apparel, including eye and respiratory protection.

### *Masking*

Media blasting will leave a slightly textured surface, regardless of the media being used. Therefore, machined surfaces, bearing surfaces, threaded areas, or any other areas that would be negatively affected by this should be masked off to avoid the blasting media. Exterior threads, such as those on the back of some trim pieces, can be easily protected from blasting by covering them with a length of appropriately sized rubber hose or tubing. Other areas can be masked with heavy cardboard and masking tape, accompanied by prudent use of lower blasting pressure and a careful aim.

### *Media Selection*

Various materials other than sand are available for use in media blasting. Media used for blasting must be compatible with the material upon which it is being used, or you will find that you have done yourself a disservice. If the blasting media is harder than the surface being prepped, you will do more harm than good by hurling hard objects at it. A large volume of softer material passing by the surface is a more appropriate way of freeing the surface of unwanted material, such as paint or rust. These softer materials typically include silica sand, aluminum oxide, plastic media, or walnut shells. You should avoid using steel shot media or coarse river sand. When choosing a blasting media, you must remember that any scratches or abrasions that you put into the metal while cleaning will also need to be taken out. An aggressive media will no doubt remove paint and other finishes faster, but if the media is harder than the material that is being blasted, you will get to a point where you are creating more work for yourself.

Although it is not the best media for any particular cleanup project, glass bead blasting does a good job on most any surface. For removing paint and rust from steel, aluminum oxide is a good choice. A little more expensive is plastic media, which is best for stripping paint from metal, as it does not get as hot and cause warping. Even though it is more expensive, aluminum shot is best for cleaning soft metals, such as aluminum, die-cast, or brass.

The following chart gives media and air pressure settings recommendations for various blasting projects.

### Media Blasting Reference

| Blast Media | Surface Material | Type of Blasting | Recommended Air Pressure (psi) |
| --- | --- | --- | --- |
| Glass Bead | Aluminum, Brass, Die-cast | Cleaning | 60 |
| Aluminum Oxide | Steel | Removal of rust and paint. To increase adhesion of paint or powder coating. | 80–90 |
| Silicon Carbide | Steel | Preparation for welding | 80–90 |
| Walnut Shells | Engine/transmission assemblies | Cleaning | 80 |
| Plastic Media | Sheet Metal | Paint removal | 30–90 |
| Aluminum Shot | Aluminum, Brass, Die-cast | Cleaning | 80–90 |

Various media can be used for removing paint and rust from the various materials found in a hot rod, but some media work better than others. This chart gives the best choices for the materials shown.

### Media Removal

In addition to flaking rust, paint, or anything else that may have been on the part prior to blasting, you must also remove all of the blasting media when the blasting is finished. Much of the media used for blasting is recycled and used again; even if your parts were not oily or greasy, previously blasted parts may have been and those residues may now be on your parts. Any oil that is present on your parts will cause primer or paint adhesion problems, so it is imperative that all parts be cleaned thoroughly after being media blasted.

A drawback to media blasting is that the media can be difficult to remove from confined areas. It is fine for use on a simple two-sided surface, such as a fender. However, on a door shell, pickup truck cab, or passenger car body, some of the media will collect between panels and will be difficult to remove. Several passes with a strong shop vacuum, shots of compressed air to dislodge media stuck in crevices, and more vacuuming will be necessary to remove all of the media.

### Grinding/Sanding

If you simply need to remove surface rust or paint in a localized area on a few parts, an electric or pneumatic orbital sander will do the job quite sufficiently. By using a 36-grit sanding disc, you can remove old body filler or previous layers of paint and primer very quickly and exactly where you desire. This can often be done in less time than it would take to load the parts in a truck to transport them to a media blaster or chemical-stripping facility.

However, you should limit this method to relatively small localized areas, as the process can produce significant heat that will cause otherwise perfectly good panels to begin warping. Although it undoubtedly has been done in the past, you should not attempt to remove paint from an entire vehicle by using a sander. You would no doubt create enough heat to cause significant warping, you would go through several sanding discs, and this process would take too much time. Even when being used to strip a small area, the sanding disc should be moved around, as opposed to simply holding the sander in one spot. You should also hold the sander so that the sanding disc flexes slightly, rather than flat against the surfaces being stripped.

Regardless of the method used to remove paint and rust from a metal surface, you must take significant steps to protect the surface or rust will return. It should be noted that an application of primer-surfacer or even epoxy primer alone will not prevent rust from reoccurring. Proper application of paint to all metal surfaces, along with regular washing and applications of wax, is the best defense. Until paint is applied, an application of epoxy primer is the most effective rust deterrent. Simply applying primer-surfacer will not prevent rust as primer-surfacer is actually porous and therefore soaks up moisture, which is the primary cause of rust.

## PATCH PANELS AND PANEL REPLACEMENT

Oftentimes, rust or collision damage will be significant enough to make replacing the panel or a portion thereof the only feasible repair. An ever-increasing aftermarket supply of patch panels is available, so you should consult your local hot rod parts supplier(s) before making something from scratch. Unless the area that you need to patch is perfectly straight, it will be easier to weld in a commercial patch panel that already has the necessary body lines and curves in it than to form your own. However, you should be aware that some patch panels will require a certain amount of hammer-induced finesse to fit as they should, yet this is still usually easier than creating your own patch panel. An option to consider if no reproduction patch panels are available is to find an original panel that is in better condition than what you already have. It might be that the repair and prep work to a second panel would simply be easier or you may have to use the best portions of each piece to assemble an acceptable panel.

This '40 Ford fender is an original Ford piece that had been severely modified in an attempt to use some custom headlights. That idea was scrapped by the current owner, so the necessary portion of a second set of fenders have been grafted into the originals.

When work is finished on this '40 Ford, you will never know that these fenders once had Mercedes headlights installed in them. The previously butchered headlight areas have been restored by grafting in the necessary sheet metal from a pair of original fenders that were damaged elsewhere but still had good headlight areas.

## Fenders

Being located at the outer extremities of our vehicles, fenders are most vulnerable to collision damage and therefore shoddy repair. After all, vintage fenders of hot rods are all now past 60 years old. Over the years, lead and poor quality plastic body fillers are likely to have been used to make repairs to damaged fenders.

While reproduction fenders are readily available for a wide variety of hot rods, they are often pricey and are not available for every hot rod project. If the damage requires more than simple hammer and dolly work, it may be necessary to make your own patch panels. Depending on the size of the patch panel that is required, you may need to use an English wheel to re-create a sweeping curve, or a birdshot-filled beater bag and a hammer to form a more intricate shape. Regardless of how you make the patch panel, the area to be replaced can be cut out and your custom patch welded in its place.

This fender probably is not pristine beneath the surface rust, or it would not be on the scrap pile. However, it might be better than what comes with the next project, so hang onto it as long as you have room to store it.

Begin by bolting the fender being repaired onto the vehicle or onto a jig so that any repair work does not distort the fender. Then cut out the areas that must be replaced. Fabricate the necessary patch panels and tack weld them in place. MIG welding will work, but TIG welding will require less follow-up. Use a body hammer and dolly to get the repaired area as straight as possible. Spray a guide coat of rattle can enamel onto the entire repaired surface, and then block sand it to reveal high and low spots. Follow this up with more hammer and dolly work as required, and then apply a skim coat of body filler to complete the repair.

### Door Bottoms

Due to their location on the vehicle, door bottoms on vintage vehicles are a common location for the formation of rust. Over time, dust, dirt, and other debris will find their way into the window opening and collect at the inside bottom of the door, clogging drain passages in the process. As rainwater runs down the window and collects on the inside of the door, it combines with the dust and dirt to begin eating away at the sheet metal. As this happens, the bottom of the door begins to deteriorate. Since the doors are usually closed whenever anyone is working beneath the vehicle, no one sees that the bottom of the door is beginning to rust, so the process continues unimpeded.

If a replacement patch panel is available (either a reproduction piece or a section cut out of an otherwise useless door), the repair is fairly easy. The main thing to remember when ordering a replacement or scavenging a piece from a donor is to make sure that the replacement section is large enough to cover the area that must be replaced.

*Continued on page 80*

Shown is a door with its typical rust at the bottom outside. A door bottom patch would be able to save this door and at significantly less cost than a replacement door.

# PROJECT 4
# **Replacing Door Bottoms**

**W**hile replacing complete doors on hot rods is perhaps not as common as on a muscle car restoration, replacing door bottoms on the older vehicles is a common occurrence. To see how it is done, I visited Morfab Customs, where Kris Valbuena was replacing the bottoms on a pair of early GM truck doors. With a replacement door bottom, a grinder, welder, hammer, and dolly, this is a relatively easy process that can make an old door look as good as new.

**1**

The passenger side door is fairly straight, but it is beginning to split just above the lower edge from about the middle toward the front of the door.

**2**

The driver side door, on the other hand, is certainly worse for wear. There is no longer a line that defines the lower edge. A pair of door bottom replacement patches, a welder capable of welding sheet metal, and a few hours labor will have these doors looking great.

**3**

The replacement door bottoms for this project include the inner skin and a slight edge of the outer skin. With the doors' slight curve, the portion of the outer skin that must be replaced can be done with flat sheet metal. With the door lying on a worktable, Kris measured the height of the replacement panel and then measured that same amount on the inside of the door. This is where the old material will be cut off the bottom of the door. To make the line easier to see, Kris adds a strip of masking tape with its top edge along the "cut" line.

**4**

Kris then uses a die grinder with a cut-off wheel to slice through the sheet metal. Be sure to wear eye protection, as well as a long sleeve shirt, to avoid injury from the flying sparks. After cutting all the way across the door, remove the original lower panel.

**5**

This is what the lower inside of the door looks like with the rusty metal removed.

**6**

Kris checks the fit of the replacement panel with the old door. It needs to be checked for length, height, and depth. You should check the length to make sure that any character lines line up correctly. Height needs to be checked to verify that the door will fit in the door opening correctly. Depth (door thickness) should be verified to make sure that there is not a horizontal gap remaining when the two are welded together. If the door thickness is different, the edge can be split and material added or removed to make the two pieces the same thickness.

**7**

Kris then uses a sanding disc to rough up the metal slightly to allow for good weld penetration. None of the metal needs to be ground away, just the paint.

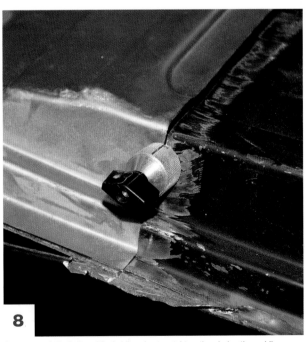

**8**

Clamps specially designed for holding sheet metal together during the welding process are applied to the door and the replacement panel. These clamps have a thin piece of T-shaped metal that slides in the gap between the two panels. The T-shaped metal then rotates 90 degrees when the knob is tightened, pulling the sheetmetal panels together. After the panels are tack welded together, the clamps can be loosened and removed.

**9**

Alignment is checked again. The door edge doesn't line up close enough for Kris' taste, so he uses a screwdriver to pry the replacement panel outward slightly.

**10**

The stamping had a nice recess in it that looked very much like the original, but it simply did not line up correctly with the original door. With a little bit of hammer and dolly work, Kris had the body line moved to where it needed to be. Although patch panels save lots of time when doing bodywork, they are usually not a simple "weld it in and be done" proposition.

**11**

Now that the body lines are aligned, Kris clamps the replacement panel back in place.

**12**

A slight misalignment is found near the far side of the door, so Kris uses a small screwdriver to pry the door upward; then he hammers the high spot down slightly to achieve the perfect fit.

**13**

Using a TIG welder to achieve the most precise weld and the least amount of buildup, Kris tack welds the replacement panel to the original door.

**14**

The inner portion of the door panel is now fully welded in place.

**15**

Although the weld seam is minimal, Kris still makes a habit of smoothing it up with a sanding disc.

**16**

Sanding this relatively narrow weld bead is a perfect example of where a die grinder with a 90-degree head comes in handy. A full-size body sander (7- or 9-inch) would be overkill for this application, while a standard die grinder would be awkward to use in this situation.

**17**

After cleanup with the sanding disc, the weld all but disappears.

**18**

A skim coat of body filler and a coat of primer will make this old truck door look brand new.

*Continued from page 75*

### Floors

Consisting of two or more sheetmetal panels that are flanged to overlap, floor panels are prime targets for the formation of rust. Although seam sealer from the factory will prevent moisture and dirt from finding their way between these panels for a while, being constantly exposed to the elements, this seam sealer will deteriorate eventually. When it does, road salt from winter weather, rainwater, dirt, and debris will work their way between these panels and start the rust formulation process.

Replacement floor panels are becoming more widely available, thanks to a growing automotive aftermarket. However, even if replacement panels are not available for your particular vehicle, making new panels is not beyond the abilities of an amateur body man who has some basic metal working skills. Granted, some vehicle floors are considerably simpler than others. Since you will be beginning with flat sheet metal, you should use poster board or other similar material to make patterns, and then transfer the patterns to the sheet metal before cutting the metal. Rather than attempt to cut out one piece of sheet metal and form it to fit the entire area, consider using several pieces, each with simpler bends, and then weld them together as required.

You should also utilize some method to stiffen the floor so that it does not "oil can" or flex. This can be done by rolling a series of beads into the floor panel with a bead roller or by attaching sheetmetal hat channels to the underside. A hat channel, as the name implies, looks like a line-drawing hat in profile. It's a piece of sheet metal with a flat flange on each side and a three-sided channel in the middle. The flat flange, or "brim," on either side can be welded or riveted to the floor to strengthen it and limit flexing.

### Stock Height

When replacing the stock floor in a steel-bodied hot rod, the most difficult portion of the task may be removing the old floor. To prevent the body from collapsing upon itself when the floor is removed, you should temporarily tack weld strap steel, angle iron, bar stock, or even electrical conduit from side to side in several places. It is best to do this prior to removing the body from the chassis, if possible.

Some vehicles, such as Model A Fords, have steel cross sills and subrails that are separate pieces from the actual floor. For vehicles that have these structural members below the floor, the new floor can be installed by plug welding the floor panels to them. This prevents the floor from being too low in the vehicle. For vehicles that do not have cross sills, the outer edge of the body typically has a flange on the lower inside to which the floor is attached. In this instance, mount the body to the chassis in the intended location using the stock body mounts. Install replacement floor panels (reproduction or custom made) by beginning in the middle of the vehicle and working your way outward, trimming as necessary to fit the contours of the inner body. When the floor panels are in position, tack weld them together to make sure everything fits as it should. Then finish welding the floor panels together. Be sure to temporarily support the new floor from underneath when tack welding it together and to the body.

## Channeling

As most hot rods sit atop the frame rails, channeling lowers the profile of the vehicle by raising the floor in relation to the body, dropping the body down over the frame rails in the process. This can be done in varying amounts with great aesthetic results, but you must be aware from the beginning that fitment of everything forward of the firewall will be affected as well. Since the firewall will be lower, the grille shell must be also, and therefore the radiator must be made shorter (lower). Additionally, passenger legroom will be decreased by the amount of the channeling.

If the floor is still in the vehicle, you can refrain from installing the temporary bracing as mentioned previously. Determine how much you want to channel the vehicle and mark that amount above the existing floor on the insides of the body. Bend a piece of sheet metal into an angle, and then use a shrinker/stretcher to form this angle to the contour of the inside of the body. Repeat for the opposite side of the vehicle and then weld one leg of the angle to the inside of the body, with the other leg providing a flange. Now tack weld a series of supports made of square (or rectangular) steel tubing from one side to the other across the inside of the body to serve as stringers for the new floor, which will be installed later. These stringers will ultimately support the weight of the body, so they must be strong, but you can come back later to fully weld them.

A prudent step now would be to support the body (on or off the chassis) and cut out as much of the original floor as necessary to allow the body to slip down over the frame rails. With this portion of the floor removed and the body back on the chassis, step back away from the vehicle and see if the new stance is what you truly desire. If it is not, take the body back off the chassis, rebuild the floor as necessary, and remove the stringers and angle strips that you had just recently tack welded in place. If the new stance is along the lines of what you expected and what you want, you can finish welding the stringers in place, weld in the new floor, and remove the old floor. Whether you remove the old floor before or after installing the new floor is up to you.

While most channel jobs take the body straight down, some intentionally channel more at the front than at the back, providing an aggressive look in the process. Whether you do a straight channel or a wedge channel, you must remember to install the stringers so that they are on a uniform slope (in other words no high or low spots). Also remember that the more you lower the cowl, the more you will be required to lower the top of the grille shell to keep the hood from sloping backward. Of course, if you are not using a hood and you completely fill the engine compartment with a mountain motor, it may not be readily apparent if the grille shell is slightly higher than the cowl.

Now if you really want the channeled look, but do not want to give up legroom, you can have both with some extra

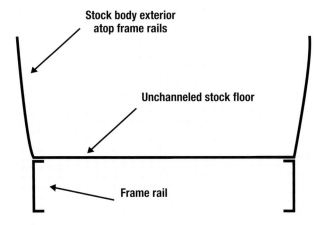

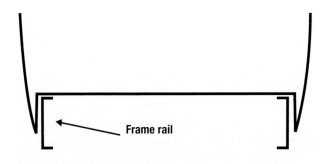

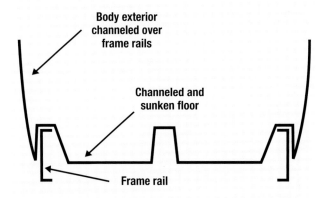

Many hot rod bodies sit atop the frame rails in their stock configuration. When the body is channeled, the floor is moved upward in relation to the body, while the body sits lower on the frame. This reduces the legroom by the amount of the channel. If desired, the advanced fabricator can rework the floor to increase the legroom and have a channeled body.

effort. This will require some well thought-out design, as well as fabrication and welding skills. The basic concept would be to drop the outer sides of the body down over the rails some predetermined amount. Instead of simply raising the floor, keeping it flat and eliminating legroom, you could channel the inner portion of the floor pans as well. This could potentially be done by using thicker sheet metal for the floor or by creating a square tube substructure to support the sheet metal. The floor still has to pass over the frame rails, but it could be sunk down between the frame rails to maintain some legroom. You just have to make sure that you do not sink the floor so much as to create a potential problem of bottoming out or having a transmission hump and driveshaft tunnel that are obnoxiously high.

### Firewalls

For Ford vehicles 1928 through 1932, the firewall simply unbolts from the body, so replacing one from those vehicles is fairly straightforward. To remove the firewall from most other vehicles, you should first tack weld a piece of bar stock across the lower edge of the body to prevent the body from squeezing together or pulling apart after the firewall is removed. Then drill out the spot welds that secure the firewall and body together and slide out the firewall. The replacement firewall can be slid into position, welded into place, and any temporary bracing removed.

This stock firewall is fairly straight, with few if any dents. However, there are several holes (either stock or added by previous owners) that will not be required for the reincarnation of this particular hot rod. The smaller holes can probably just be welded up, while the three larger ones will require a piece of sheet metal to be welded to the back side. The original cowl vent (located on top of the cowl) will most likely be retained on this vintage rod, but patch panels are available should you desire to remove the vent on your cowl.

# PROJECT 5
# Installing a Patch Panel

**P**rior to cutting away any of the old sheet metal, you must first verify that the new patch panel will span the entire area that needs to be removed. Many similar patch panels are available with varying amounts of surrounding area included and are priced accordingly. Unless the panel is in a severely complex area of the body, installation is going to be comparable, so you should use the correct size panel in the first place, rather than patch the patch panel.

Whether you can hold the patch panel up against the area where it will be installed, or you are required to measure from similar points, determine where the old metal can be cut and mark a line on the old metal. Now mark a second line approximately a half-inch inside of the first line so that you will be cutting away less material than your original estimate. You will most likely need to cut away more metal before you weld the patch panel in place, but this half-inch will provide for some fine-tuning and a bit of oops factor.

Use a plasma cutter, die grinder, reciprocating saw, or tin snips to cut away the original metal. You must be cutting some good metal so that you have solid metal to weld to. Trial fit the patch panel to verify that it fits correctly, and then mark any additional original metal that must be cut away for proper fit. Cut the remaining metal to the correct size and shape. Now clamp the patch panel in place, double-check for proper alignment, and then tack weld it in place. With the patch panel securely tacked in place, verify that the perimeter is in the correct relationship with the original edge to which it will be welded. That is, make sure that the edge is not too high or too low. Use a hammer and dolly to work the patch panel and surrounding area to the correct contour. Now finish welding the patch panel around its perimeter, making sure to skip around to avoid heating the metal too much in any one spot. Another tip is to use your air hose with an air nozzle to blow cool air on the welds to keep them from getting too hot and causing distortion.

After all welding is completed and the metal allowed to cool to room temperature, clean the patched area with wax and grease remover, and apply epoxy primer. Follow with primer-surfacer and block sanding as necessary prior to sealer and top coats.

**1**

When installing a patch panel, you must first determine how much of the original panel can be/should be removed. You should be able to remove all of the damaged or rusted area, but not so much that the patch doesn't cover the entire area. On this quarter panel of a Ford Model A coupe, Keith Moritz fits the panel in place, then scribes or marks a line on the original panel. Ultimately, the original panel below this line will be removed.

**2**

Since the patch panel will be welded into place, the existing paint (and body filler in this case) are ground away with a grinder. Approximately 2 inches of bare metal should surround the area to be welded.

**3**

Before cutting the original metal away, Keith double-checks the fit of the patch panel and then scribes the line along which the cut will be made. It pays to double-check the fit of the patch panel before cutting anything, so that you don't have to weld in a patch between the original metal and the patch panel you are installing. It is much easier to make a second cut than to add more material.

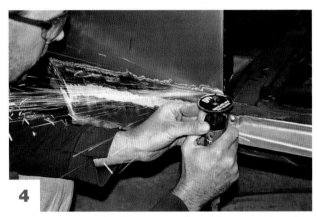

**4**

In this case, Keith uses a die grinder with a cut-off wheel; however, a reciprocating saw or plasma arc cutter may be used to cut away the original panel. Accurately fitting the panel, marking the cut accordingly, and then accurately cutting out the panel will provide a better patch panel installation.

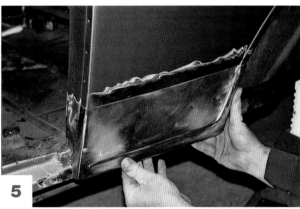

**5**

This is the opposite side of the same vehicle, receiving the same patch panel installation. After the defective area is cut out, the new patch panel is fit into position making sure that any and all body lines are correctly aligned. The patch can then be clamped into place to allow for welding.

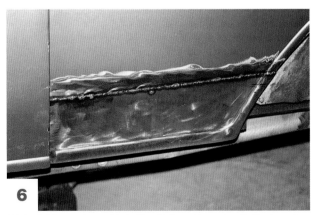

**6**

This is what the quarter panel looks like with the patch panel welded in place prior to the welds being ground down. Notice that the body lines are correctly aligned and the patch along with an area above it have been lightly ground to allow for proper adhesion of epoxy primer.

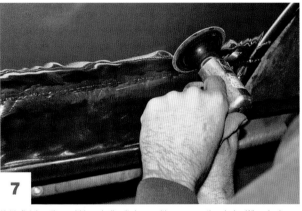

**7**

Keith finishes the weld by grinding it down with a pneumatic grinder. When he is finished, it will be virtually impossible to tell where the seam is located. Depending on the condition of the surrounding area and the amount of previous bodywork, some body filler and primer-surfacer may be necessary to fully prep this area for paint.

# FILLING HOLES AND DINGS

Occasionally, you will have the need to fill small holes, tears, or gaps in sheet metal. These are commonly left over from previous installations of trim, accessories, or routing for wiring. Larger examples are often caused by rust, collision, or a variety of modifications that may have been attempted by previous owners. The best method of repair depends largely on the size of the hole or gap to be filled. Whatever may be the cause of the hole that needs to be filled, this applies to areas where no reproduction patch panels are available.

### Welding Shut

As long as the hole is no larger than about a quarter-inch in diameter, you can simply weld the hole shut. After it is welded closed, grind off any excess that is above the surface and then sand smooth. If the hole is larger than a quarter-inch in diameter, cut out a small piece of sheet metal, place it over the hole from the back side, and then weld in place. Again, grind off any excess weld and then sand smooth.

### Using a Filler Patch

If the area to be closed is larger than a postage stamp, you will need to determine if the original sheet metal was flat in that area or if it was curved or had any body lines. The basic shape will need to be reproduced, prior to welding the custom patch panel in place. If the panel simply needs

Shown are two small holes that have simply been welded shut, without any sheet metal being welded in behind them. The weld has been ground smooth and will be finished with a skim coat of body filler.

to be curved slightly, you can gently wrap the sheet metal around a welding gas tank until the approximate shape is reproduced. If the panel requires significant curve, it may be necessary to use an English wheel to fabricate the desired patch panel. Body hammers and dollies will be required to re-create body lines.

This is the location of the stock taillights on a '40 Ford sedan. Different taillights will be used in a different location, so the stock hole needed to be filled. The adjacent area was scuffed up with a grinder and then a filler piece of metal cut to the required shape and contour. It was then welded in place and the welds ground smooth. A skim coat of filler will remove all traces of the hole.

# PROJECT 6
# Filling Holes

**M**ost likely any original steel body that you build into a hot rod will have at least one and potentially several small holes in the sheet metal that you no longer want or need. If the hole is approximately a quarter-inch in diameter or smaller, it can be welded up without the use of a patch. Most holes that are any larger should be filled by welding a sheet metal patch in place. Mike "New Guy" Schafer at Morfab Customs showed me how it is done.

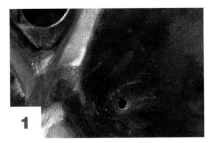

**1** While the big hole at the top of the photo is necessary, the small hole near the middle of the photo is not required, so it will be filled. Many amateur body men would try to fill this hole with body filler and be done with it, but that remedy has a great potential for falling out eventually.

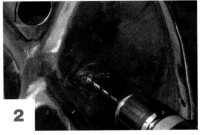

**2** This surface is fairly clean, but if it wasn't, Mike would use a grinder or wire brush to clean away any paint, primer, or rust. An important and sometimes overlooked step is to remove any rust at the very edge of the hole. This rust could potentially cause problems with the weld, so Mike uses the next size larger drill bit to clean the existing hole. This gets rid of any rust or other residue.

**3** Using a TIG welder, Mike then carefully heats up the sheet metal around the hole . . .

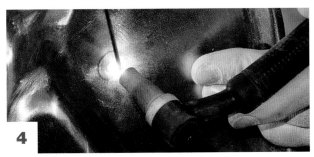

**4** . . . and then inserts the filler rod at just the right time.

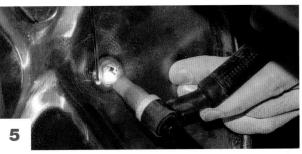

**5** By applying heat as required, Mike can keep the welding puddle fluid as he adds filler metal to completely fill the hole.

**6** Still glowing red hot, we can see that the hole is now filled.

**7** Mike now uses a 50- or 80-grit sanding disc on an angle-head grinder to smooth the excess weld. Depending on how careful Mike is with the grinder, this repair may not need any filler, but there is no shame in using a skim coat of filler to feather repairs like this.

# PROJECT 7
# Filling Bigger Holes

**N**ot all holes to be filled are smaller than ¼-inch in diameter or have commercial patch panels available to fill them. When you find yourself in this situation, you may need to make your own custom patch panel and then weld it in place. After cutting out the rust, and squaring up the hole, Kris Valbuena at Morfab Customs whipped up this little patch panel. Keith "the Boss" Moritz welded it in place.

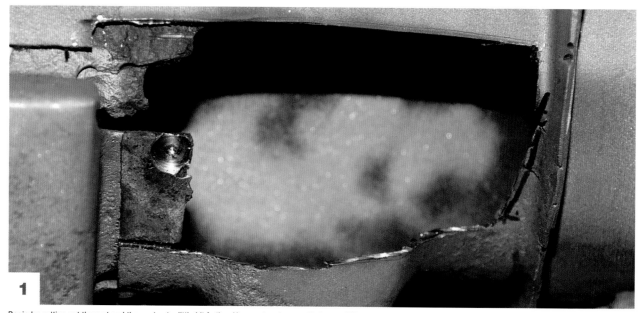

**1** Begin by cutting out the rust and then cut out a little bit farther. You must make sure that you will be welding the patch panel to solid metal. The shape of the hole and the patch panel to fill it are up to you, but making the hole a shape that is easy for you to reproduce in sheet metal should be a goal.

**2** Using a couple of small, but incredibly strong magnets, the custom patch panel is temporarily held in place. Magnets are the convenient way to hold this patch panel in place prior to welding.

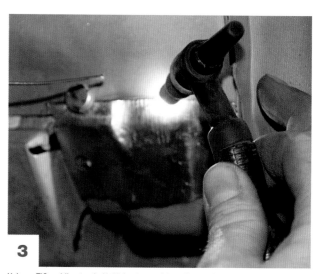

**3** Using a TIG welding torch, Keith heats a point on the patch panel and the surrounding area . . .

**4**

. . . and then hand feeds the filler rod when the temperature is just right. The TIG welder uses a higher temperature than a MIG but concentrates that higher temperature in a tighter area. Keeping the heat confined to a smaller area minimizes distortion.

**5**

The heat circle of the weld just made is approximately ½-inch in diameter, testifying to the controllability of the TIG welder.

**6**

After adding another tack weld, Keith flattens the weld by hitting it with a body hammer while the metal is still red hot.

**7**

The second bead from the right on the top is the one in the previous photo after being flattened while still hot. Not only does TIG welding contain the heat within a small area, but it can be done in a way that minimizes follow-up bodywork later.

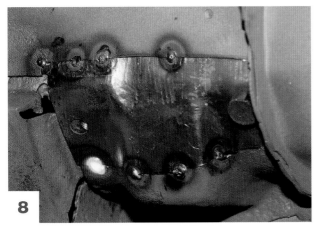

**8**

As indicated by the glow, the metal is extremely hot, yet it is contained within a very small area.

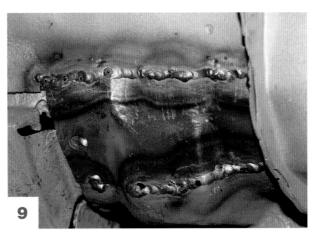

**9**

When this hot rod is finished, this area will most likely never be seen, yet it will most likely be cleaned up some anyway.

# Chapter 4
# Reproduction Bodies

**L**ove 'em or hate 'em, there is no denying that reproduction bodies make hot rodding more affordable for the masses. Hot rodding pre-1948 automobiles would not be nearly as popular as it is today, as there simply would not be as many around, if not for reproductions. What began as fiberglass reproductions of a select few Ford models has now grown to include several models of Fords and various Chevrolet, Willys, and Mopar products. A few roadster models have been re-created in steel for a few years now, while '32 and '33–'34 Ford coupes are now a reality, with the '40 Ford coupe being in development.

## PROPER STORAGE OF FIBERGLASS BODIES

Due to the very nature of fiberglass resin, not only complete bodies, but any parts made of fiberglass must be allowed to fully cure. Until all of the solvents in the fiberglass have evaporated, the fiberglass parts will move around slightly unless they are somehow restrained. Some body manufacturers allow sufficient cure time prior to shipping their parts, while other companies ship product sooner. If the body is not fully cured when you take delivery of it, you must take appropriate

action if you intend to obtain the best fit and finish possible on your new body.

At the very minimum, you should plan to have on hand the perimeter frame that you will be mounting your new body on when you receive the body. The ideal situation would be to have the entire chassis, suspension, wheels, and tires already assembled, so that the body can be mounted correctly from the beginning. If you do not have the frame that you intend to use, set the body on a flat floor and make sure that nothing is resting against it. You may not notice it right away, but if something was leaned against the body before it has fully cured, it will very likely cause more work for you in the future when you realize the body is not as straight as it could have been.

If you do have the chassis that you intend to use, mount the body on the chassis, fit everything (fenders, doors, deck lid, hood, etc.), so that gaps are consistent. Sand all of the fiberglass parts with 80- or 120-grit sandpaper to break the gel coat, and then allow the chassis and body to sit in the sun as long as possible. Opening the gel coat and allowing the body to bask in the sun will help to completely cure the body. You should not bother using any grit finer than 120 at

When you plunk down the big money for a fiberglass body, you should be sure to get it mounted on the chassis as soon as possible. Before fiberglass fully cures, it is can easily be distorted if something is leaning against it or otherwise putting a force onto it. This is part of the reason body manufacturers like to sell chassis/body packages together.

this point, as you simply want to break the seal and allow the fiberglass to breathe. Block sanding should be done only after the fiberglass has cured completely.

## Mounting the Body on the Chassis

When mounting the body on the chassis for the first time, you will ideally have all four wheels and tires installed already. Besides the safety aspect of not being able to inadvertently knock the chassis off the jack stands while in the process of lowering the body onto the chassis, having at least the back tires in place will help to ensure that the body is aesthetically in the proper position. Most stock and reproduction chassis have mounting holes for the radiator, so begin by installing the radiator in the correct location and with the correct hardware. Now is not the time to "just sorta wing it"—make sure that the radiator is where it belongs and is not going to be moving around while installing the body. Now mount the grille shell and then measure the hood to determine how far back the cowl should be from the grille shell. Mark the top of the frame rails at the location of the bottom of the cowl. You can use a marker or tape, which ever works best for you. Presumably you have not yet painted the chassis if this is indeed the first time the body has been mounted.

Now round up all of your buddies so that you have enough folks to lift, move, and carefully set the body down on the frame rails. Be very careful so that no one gets injured. A couple of extra people would be good to have on hand to act as spotters to make sure that nothing is getting caught, damaged, or otherwise mishandled. Position the body so that the bottom of the cowl is aligned with the marks you made on the tops of the frame rails previously. Take some measurements from various locations on both sides of the vehicle to make sure that the body is square on the chassis. Now step back away from the vehicle and check to make sure that the rear tires are centered in the rear wheel wells. If they are not, now is the time to make adjustments. Is the rear axle in the correct location and square in the chassis? Is the cowl aligned with the marks on the frame rails? If the rear axle is indeed in the correct location, you can slide the body forward or backward slightly to align the wheel well and the tire, as the radiator and grille shell can be adjusted slightly as well to make the hood fit.

Original and reproduction chassis have body mount holes in them already, while most reproduction bodies do not. As a minimum, there should be body mount holes located at the front of the cowl, at the front of each door opening, at the rear of each door opening, and near the rear of the body on each side of the body. So now, the task is to determine where the mounting holes need to be in the body so that they align with the holes in the frame rails. With the body sitting in what you have determined to be the correct location, make the necessary reference marks on the body and chassis so that you can align them again. There are multiple methods for determining hole locations, but I will discuss two of them.

One method requires you to first remove the body from the chassis, determine what size bolts are required for your particular chassis (usually $5/16$ or $3/8$ inch), and then collect studs or threaded rod in the appropriate size, one for each body mount hole. Using a die grinder or bench grinder, grind a point on one end of each stud. Then thread a stud into each body mount hole in the chassis with the point up. If you are going to be doing this on a regular basis, you can tack weld an appropriately sized nut in about the middle of the stud so that you can get a wrench on it should it decide to get stubborn when you are ready to remove it. With the studs pointing upward no more than an inch above the tops of the frame rails, now set the body back in place, using the alignment marks that you made previously. With a whack of a rubber mallet on the inside of the body at each body mount location, the pointed studs will transfer the body mount hole location to the bottom of the body. Now remove the body and safely support it so that you have access to the bottom side of it. If necessary, simply drill a pilot hole from underneath, and then drill the correct size hole from the inside of the body. Be sure to drill the body mount holes slightly larger than the body mount bolts, to better facilitate installation of the body at a later time. Remember to remove the studs prior to reinstalling the body again.

A second method of determining body mount locations calls for some work to be done prior to installing the body the first time (in order to avoid installing and removing another time). Place a piece of masking tape on the inside face of the frame rail near the top. Using a square (or a careful eye), mark a vertical line on the tape indicating the center of the mounting hole. Then note the actual distance from the top inside frame rail outward to the center of the hole. After doing this for all the mounting holes, position the body on the chassis. When it is determined to be in the correct location as discussed previously, clamp the body to the frame if possible. Now crawl beneath the vehicle again with masking tape and marker in hand. At each mounting hole location, place a piece of masking tape onto the bottom of the body, so that the outer edge of the tape is aligned with the top, inner frame rail. On the tape, mark the centerline of the hole (vertical line previously) and the distance outward toward the center of the hole (should be approximately half the width of the frame rail). When this is done for all body mount locations, the body can be removed and the holes drilled in accordance with the notes made on the bottom of the body.

With the body mount holes drilled and the body sitting on the chassis, you should now be able to at least match the holes between the body and the chassis. If necessary, enlarge the holes in the body slightly. Use stainless steel fender washers (they are thinner and stronger than regular fender washers), an external star washer (it will bite into the body somewhat), and the correct diameter bolt at each location to secure the body to the chassis. Be sure to start all of the bolts before fully tightening any of them.

As this very slick, highboy deuce coupe demonstrates, fenders are optional. You should install fenders if you like them, but the time saved by not fitting, sanding, and painting fenders on this hot rod went into getting the rest of the body arrow straight.

## Installing Fenders

After your first few attempts at installing fenders, you will have a better understanding of why many hot rodders choose to forego using them at all. Not that the process of installing fenders is really difficult, but to get them installed the same on both sides will leave you scratching your head more times than not.

Whether the body is made of steel or fiberglass, proper assembly begins with mounting the body on the chassis. This was discussed earlier, but some key points are worth repeating. Make sure that the rear wheel wells are properly aligned with the rear axle. Read that as having the wheels and tires mounted to the axle, the fenders (if being used) are mounted, and then stand back to make sure that it *looks* right. While I don't recall ever seeing anyone walking around a rod run measuring tires to see if they are dimensionally centered in the wheel well, I have seen, heard, and admittedly even commented myself on wheels and tires not being center in the wheel well. This is just one of those things where numbers do not matter quite as much as looking correct does. Knowing those situations in advance will ultimately make you a better hot rod builder.

Be sure to use fender washers to distribute pressure evenly over a larger area than just the head of mounting bolts. Another point worth remembering is to use stainless steel fender washers as shims, instead of the hardware store variety. The stainless steel fender washers are more consistent in thickness, are stronger, and will not rust.

Same basic car with fenders . . . four fenders and two running boards are the only major differences, but what a difference it makes.

While 1934 and earlier Ford passenger cars look good with or without the fenders, the later fat-fendered cars, such as this '36 sedan, just would not look right without fenders. You can also see that the later 1930s cars have enormous front fenders that would require an extra set of hands when installing.

Another bad habit that many hot rodders get into when trying to rush the assembly process is to paint all of the components prior to assembling them for the first time. Sadly, this usually ends up in frustration, as you will undoubtedly chip and scratch paint. Also, you will most likely find out during assembly that at least a few of the parts need to be modified to fit properly. It is certainly acceptable to prime parts (especially steel parts) before assembly, but they should not be painted. Assemble the vehicle as completely as possible prior to painting any parts, even if that means putting it together several times. This will help to ensure that all the parts fit together correctly and do not attempt to occupy the same points in space at the same time. Plus, you will be more likely to trim or grind a part as required to make it fit properly if it is still in bare steel or primer. If the piece that doesn't fit quite right is already painted, you will find yourself less likely to take a grinder to it, even though that might be exactly what is required.

For any vehicle that uses a running board splash apron between the body and the actual running board (such as a Ford Model A), you should know that the splash apron gets mounted prior to the body. You should also determine if your specific vehicle uses frame mounting pads, blocks, or any other type of spacer between the chassis and the body. The following fender and running board installation instructions are very generic, so they may not pertain to your vehicle, but they are intended to give you a place to start.

them yet, as they may need to move slightly. Now clamp the running boards in place. Mount the front splash apron to the frame horns. Bolt the front fender supports to the frame. Now clamp the front fenders to the frame, running boards, and fender brackets. If used, adjust the running boards' splash aprons to fit properly with the front fenders and running boards. Clamp the aprons to the front fenders. If a headlight bar is going to be used, install it now, using shims between the fenders and fender brackets if necessary to gain proper fit. Check the fit of fenders, running boards, and related components, and adjust as necessary to obtain correct fit. Drill necessary mounting holes and secure all components as required.

A simple, but effective method of securing and opening the hood on many vehicles is to use the stock hardware. The latches in this case mount to the fenders, but could mount to the frame rails on a fenderless car. You are correct—the hood does not fit on this deuce as well as it could.

On Ford Model A vehicles, the running board splash aprons mount between the frame and the body. The front fenders mount between the frame and the hood. So, install the running board splash aprons and the front fenders before installing the body, and then install the hood and the rear fenders.

Mount the running board brackets to the frame and verify that the brackets are at a right angle to the chassis centerline. If your vehicle uses splash aprons as mentioned previously, clamp them in place at this time. Do not bolt

When using fenders on early hot rods, lots of fitment is required to get everything lined up correctly, as several parts are intertwined. Even a simple task like drilling the headlight bar mounting holes in the front fenders must be done with care and forethought.

Much of the body-related work has to do with making everything fit within the necessary confines. In the engine compartment, that includes the engine, exhaust, steering, and brake master cylinder/reservoirs.

On the other end of the hot rod, mounting points for deck lid lifts and hinges provide multiple opportunities to fabricate neat brackets.

## Baking Fiberglass Bodies

If you live in a cooler climate or simply do not want to wait for the resin to fully cure in your new glass prior to doing final bodywork, there is an option. If there is a decent body shop in the area that has a spray booth equipped with heating capabilities, the body can be baked. You should still mount the body on the chassis, fit everything, and scuff the gel coat with 80- to 120-grit sandpaper. Move the chassis and body into the booth, then adjust the room temperature to between 200 and 300 degrees Fahrenheit for about two hours. After the heat is turned back down to its normal setting, leave the chassis and body in place until the ambient temperature stabilizes. This process will fully cure the fiberglass, making it ready for you to block sand and finish at your convenience.

Be willing to work around the body shop's schedule and be prepared to pay for the use of their spray booth, as it will be tied up for the biggest part of a day. While it may not seem like the shop is doing much, using their oven will make a difference in the overall quality of your bodywork and therefore your hot rod project.

## Bracing

Most reproduction hot rod bodies have an adequate amount of bracing installed before they leave the manufacturer. However, you must watch out for some bodies that are designed as "race only," as they are not designed for street use and do not have sufficient bracing. If you are shopping on the Internet or printed catalogs for reproduction bodies, most companies will quickly point out the difference between "race only" and "suitable for street use" products. The problem usually occurs when a lightweight, race body is found for sale at a swap meet by an unsuspecting novice who does not realize the difference. You can fabricate and install your own bracing to one of these lightweight bodies, but you will most likely minimize any swap meet savings in the process.

### Steel

Contemporary practice has most hot rod body manufacturers bracing their bodies with steel, or steel combined with wood, rather than wood alone. At the minimum, some sort of steel support is commonly found in the firewall /cowl area to provide a substantial support for the steering column. Additional steel tubing is commonly found around the door opening and the back of the passenger area.

### Wood

As automobiles originally manufactured in the 1930s and 1940s used wood for structural support, it was common for early reproductions of these bodies to utilize wood for reinforcement as well. More body manufacturers are now using steel tubing as reinforcement where possible, as it is typically easier to work with. The main use of wooden bracing now is in doors and around windows. Wood can be shaped to compound curves to fit into these areas, and

For orientation, this is the passenger side interior of a '34 Ford coupe. The steel brace at the top of the photo would be located behind the top portion of the dash. A square tubular steel brace runs upward from the floor and then turns horizontally along the top of the cowl and then back down to the floor. Another square tube runs upward from the floor at the front of the door with a piece of flat steel plate connecting the two steel tubes.

Wood is used to support the exterior door edge, as it can be easily shaped to fit the door precisely. Lots of metal goes into a hot rod, but there are still some good applications for wood, too.

it provides a workable surface for the upholsterer to attach upholstery panels.

## Protecting Steel Bodies

Just like any other steel body, original, reproduction, hot rod, or OEM, a reproduction steel hot rod body requires protection from the elements, or it will begin to rust sooner or later. Most reproduction sheetmetal parts have epoxy primer applied prior to shipment, or they have nothing at all. The following information will also apply to other sheetmetal items, such as fenders, door, deck lids, and hoods.

Parts that are bare sheet metal should be treated to a couple coats of epoxy primer. Begin by cleaning the parts with wax and grease remover, then scuff slightly with 200-grit sandpaper. Blow away any dust and sanding grit with an air nozzle, clean with wax and grease remover again, and then apply epoxy primer according to the directions for the particular epoxy primer you are using. Be sure to avoid using primer surfacer or any primer that does not use a catalyst at this point, as those products will actually attract moisture and cause rust, rather than repel it.

If the parts already have primer on them, you should inspect them to verify that there are no scratches, dents, or any other occurrences where the epoxy primer has been penetrated. If there are any places where bare metal is showing, clean with wax and grease remover, and then scuff and prime the affected area as above.

Prior to welding on the primed sheet metal, the area will need to be ground free of any and all primer, but primer is much easier to grind off than rust. Epoxy primer will also increase adhesion of any subsequent applications of body filler, should they be necessary.

In the back of the same coupe body we see lots of square steel tubing supporting the lower portion of the passenger compartment. Hardwood (usually oak, but not always) that is more easily shaped to fit around the door and window openings also provides sufficient mounting locations for upholstery panels.

A guess would be that someone began using paint stripper on the door and rear quarter of this Ford B400 and then quit before doing the cowl. Regardless of what was done, it looks like this body and frame have been stored in a dry climate for quite some time. Having the body chemically dipped or media blasted and then coating it with epoxy primer would have this vintage tin looking great in no time.

### Rust Proofing

Vintage tin is still out there and reproduction bodies made of steel are now becoming more common. Regardless of the age of the steel in an automobile body—vintage tin, late model OEM, or new reproduction hot rod body—that steel will eventually begin to rust if not protected.

The obvious method to deter rust from forming on the largest part of a steel body is to use epoxy primer, paint, and a good coat of wax on the exterior of the body. That is all well and good, but the most likely locations of rust forming on the average hot rod are the places where it cannot be seen, just as it is on your daily driver. Whenever you wash your pride and joy or get caught in a rainstorm while cruising across Indiana, water can seep down past the door glass and collect in the inside of the door shell. A simple cure for this is to drill some small drain holes in the bottom of the door shell and then make sure that they do not get clogged. It really is not feasible to paint the inside of your door shells, but you can spray rust proofing inside of them before they are covered by interior door panels.

### Weather Stripping

Proper fitting weather stripping around doors, windows, deck lids, and hoods is one of those necessary evils that body men and painters must learn to deal with when working on hot rods. With the average price of a hot rod being right up there with that of a new car, and the median price being even higher, the owner of a hot rod does not expect (nor want) to be annoyed by road noise or leaks when driving in the rain.

Although the installation of weather stripping will take place after the body has been painted, you should give the matter some thought beforehand. A common occurrence among body men and painters is that they will spend countless hours getting door and body gaps perfect, with latches and strikers holding everything tight. Then, after weather stripping is installed, closing doors, hoods, and deck lids completely becomes a chore.

The hot rod and restoration aftermarket offers a wide variety of weather stripping, both in original styles and universal, so that no matter how mild or wild your creation may be, there are solutions to eliminating leakage into your hot rod. Many original styles are designed to fit over lips in the body molding, while others are simply glued in place. Whichever style you may use, temporarily install it before fitting panels for the best results.

By assembling the body and closing the door a few more times prior to painting the vehicle, the builder could have probably adjusted the latch or shaved a bit off the door. A little bit of checking and adjusting when in primer will prevent the paint from being chipped or worn away.

If you are gonna finish the trunk of a hot rod this nicely, it deserves to have the proper weather stripping installed to protect it from the elements. Even if the trunk is not finished, weather stripping is important to keep out moisture and prevent rust.

If not using a windshield that is glued in place, you will need correct weather stripping to avoid leaks around the windshield. Even if your hot rod never sees rain, weather stripping helps to avoid those annoying wind leaks as well.

The back and side glass requires weather stripping as well. Many types of weather stripping are available in universal styles, but better-fitting model-specific kits are available for the more popular models.

# Chapter 5
# Working with Metal

**H**aving the ability to shape metal, re-create that custom shape multiple times, and then weld it into a usable form is something that separates hacks from craftsmen in the world of hot rod fabrication. Whether the task is creating a bracket to mount some component on to another, making a simple boxing plate, or radically altering the body profile, being able to work with metal yourself, rather than paying someone to do it, will unleash your creative freedom.

## MAKING PATTERNS

Any seasoned metal fabricator will tell you to make a pattern before you touch a piece of metal. Sheet metal is expensive, while Kraft paper, poster board, and newspaper are very cheap. The best material to use for pattern making depends largely on how the pattern will be used, how often it will be used, and if it is going to be used to mock up a three-dimensional shape.

For simple patterns, such as for a bracket that will be cut from a piece of steel plate, most any type of paper will be sufficient. Simply draw the shape onto a piece of paper. What you must remember, however, is to leave sufficient material around any holes that must be drilled. While designing the part, you must also be thinking ahead about how you will actually cut it out. If you are going to be cutting the part out on a band saw, how tight a radius can you cut? The pattern

will need to reflect that. Should all of the sides be parallel and perpendicular with each other, or does one of them need to be at an angle so that the finished part is level or plumb when it is installed?

For three-dimensional patterns, such as for a floor console or master cylinder mounting bracket, use poster board, cardboard, or even Foam-Core. These materials are usually rigid enough to support their own weight and can easily be cut and trimmed with scissors or a utility knife. Once the pieces are cut to size, you can tape them together with masking tape to form three-dimensional shapes. While a lightweight pattern will not support a heavy component, it will help you visualize the metal piece you need and allow you to gauge whether what you've drawn or imagined looks as good as you'd like. Once the 3-D pattern is the correct size and shape, you can disassemble it and use its components to cut out the metal shapes you'll need for the permanent part.

For any parts that will be made on a regular basis, you can bet that fabrication shops make their patterns out of sheet metal or steel plate so that they will last indefinitely and are not subject to curling or shrinkage. Also be sure to label the patterns with a permanent marker so you know what they are. For a 3-D pattern that you will store unassembled, you may want to mark the orientation of the pieces too to remember how it all fits together.

With the shifter boot trim ring in place, it is easy to see that the hole whittled in the floor during installation and adjustment will not be covered by the boot. Additionally, I didn't want a hole that big in the floor, even if it was covered up. A decent size access panel to which the boot will be mounted will be made to allow for better access.

A pair of reference marks were made on the tranny tunnel. One centerline would have been sufficient, but the dimmer switch was in the way of that.

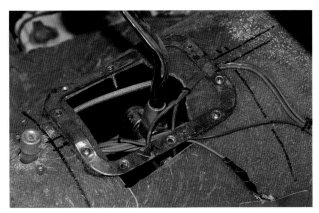

Based on the size of the aluminum panel that I had, I made parallel lines for the front and back of the access panel. These lines are perpendicular to the two reference lines.

Disregard the line that had been marked around the original location of the shifter boot trim plate. I wanted the side of the access panel to be parallel with the base of the tranny tunnel, so I measured up from the floor 2 inches at all four corners. By connecting those dots, I would have the overall size of the access panel. Parallel lines inward about an inch are where the floor can be cut out.

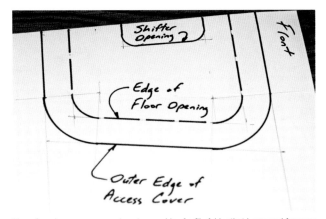

I transferred my measurements onto one side of a file folder that I procured from my office supplies. I used a drafter's radius template to round the corners, but a washer, paint can, or other suitable objects can often be found in the shop. I also marked the required opening for shifter movement.

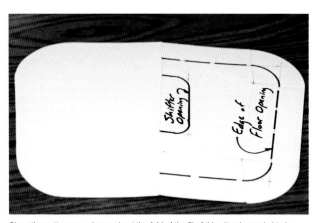

Since the pattern was drawn about the fold of the file folder, it only needed to be drawn once. The file folder can be folded closed and the shifter opening and outer shape cut out.

After the aluminum access panel was cut out, the portion of the pattern representing the edge of the floor opening was cut out and transferred to the floor. The fiberglass floor was then cut out with a saber saw.

With a good pattern, a piece of thin aluminum, some speed nuts, and bolts, I now have a much cleaner tranny access panel.

## Cutting

Almost all metalwork requires some amount of metal cutting. While the extent of your tool collection will determine your options, some tools simply will not get the job done, while other tools will be overkill for small tasks. In addition to the physical ease of cutting by the operator, you must consider accuracy, distortion, and the length of time required to do the job.

Does the pattern to be cut out require straight and curved cuts? If it is a combination of straight and curved cuts, a plasma arc cutter or a reciprocating air saw will usually work. If the piece is fairly thin, aviation snips will work, but you will not be able to cut steel plate with them. Likewise, if you are cutting a piece of thin aluminum, there is no need to use a plasma arc cutter.

For straight cuts in sheet metal, a sheetmetal shear works great. However, this only works for flat metal that has not yet been used. For making straight cuts in formed sheet metal, a cut off wheel in a die grinder would be a logical choice.

Whenever cutting metal, be sure to measure twice and cut once. Wear proper safety equipment and deburr the edges when finished cutting.

### Top Chopping Guidelines

The focal points in a top chop are the pillars supporting the roof and the doorposts, from which material is removed to set the top lower on the vehicle and reduce window height. Because A and C pillars are typically angled, rather than straight up and down, removing material there requires adding some to the top or increasing the angle on the support pillars so that the top can be reconnected to the chopped-down posts.

With the excess window area in a stock Ford Model T or Model A, the proportions of the vehicle can be greatly improved by chopping the top. Chopping the top on a '40 Ford coupe, on the other hand, usually does not improve the proportions, but that's a matter of opinion. The amount of the chop is critical as well. Three inches out of the top of a Model A is about right, while 6 inches is about right on a '51 Chevy wagon. Let moderation be your guide, as too much is usually worse than not enough. And if you're tall, take that into account, so you don't build something that doesn't fit you.

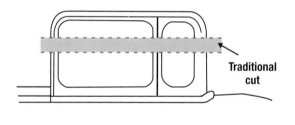

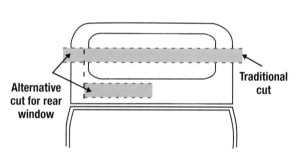

On a square-top hot rod, the cut for the top chop can be made most anywhere within the window area. To chop the top, yet retain the rear window size, the cut across the back of the car can be moved to below the rear window.

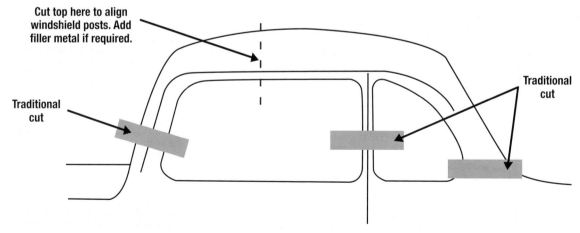

On a turret top hot rod, material is removed from the shaded areas as shown. Making the cuts as low as possible helps to keep the cuts consistent in height. Since the windshield frame is slanted, the same amount of material being cut will yield a different amount of chop. To compensate for this, the top may need to be cut above the door and material added or removed to align everything.

As shown on this vintage '32 Ford coupe, the windshield and doorposts are vertical. A relatively simple cut of material through the glass area brings the top straight down.

If you want to test your metalworking and bodywork skills, chop the top on a panel delivery. Whether a straight or wedge chop, this is a lot of sheet metal to cut and weld back together and keep straight.

A Model T Ford has straight posts that would make for a simple top chop. However, the builder of this hot rod took more out of the windshield posts to create a wedge chop, which provides a much more aggressive look for this car.

Although the sides of a '40 Ford are relatively vertical, the windshield leans back, and the back glass leans forward. To chop the top, the top must be expanded front to back to meet up with the posts, the posts leaned in, or a combination of both.

On this Advance Design Chevrolet pickup, the top has been chopped by leaning the windshield posts back. With this style of chop, the finished profile looks very natural, and therefore may go unnoticed by those not familiar with these trucks.

Shown is a stock height '47–'53 Chevy pickup truck. As you can see, the stock windshield is not quite vertical, but certainly more so than the previous truck with the chopped top and leaned back windshield posts.

For the purposes of this book, most hot rod bodies fall into one of two categories: square top or turret top. The square top hot rods are those where the window and door pillars are vertical in both profile and front/rear perspectives. These are the easiest chops to make, as the same amount of material can be removed from just about anywhere within the window and door pillars and the top drops straight down. Likewise, the interior garnish moldings can be cut in similar fashion and joined back together with relative ease. These vehicles have flat glass, so having new glass cut to fit the new opening is easily done as well.

While the smaller windows that result from a top chop are part of the reasoning for cutting the top in the first place, this does make looking out the back window difficult at times. To avoid the mail-slot rear window effect on one of these square top vehicles, (some of) the chopped material can be cut out from beneath the rear window. Extend the slice beneath the window beyond the edges of the window opening 2 or 3 inches, then cut vertically until meeting the cut made across the side windows.

Turret top vehicles, such as most passenger cars originally built in the mid-1930s up through the late 1940s, have window and door pillars that are farther apart at the bottom than at the top. In some cases, the window and door pillars slant in both profile and front/rear perspectives. With the slanted posts, the uncut top is already smaller than the area at the feet of the posts. To chop the top on one of these vehicles, you have two basic options, with both of them being required on some vehicles. One method is to cut the top in half or quarters and weld a sheetmetal strip in between the outer portions, making the top larger in the process. A second method is to put a V-shaped notch in the windshield posts and increase their angle. Generally speaking, the more the top is chopped, the more likely it will be that both methods will be required.

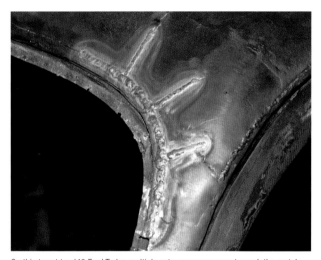

On this turret top '40 Ford Tudor, multiple cuts were necessary to work the metal together above the corner of the windshield frame after the top was chopped.

After grinding the welds smooth and adding a skim coat of body filler, that same '40 Tudor is looking much better.

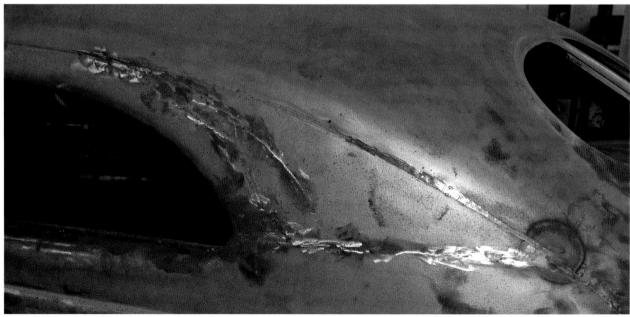

At the back of that '40 Ford Tudor, several cuts were made in the rear quarter panel as well. While these cuts stand out in bare metal, you can bet they will be smooth when finished.

When a top is chopped the correct amount, it will usually improve that vehicle's proportions. Many rodders will argue that a stock '34 Ford coupe is among the Blue Oval's best looking vehicles, but even it can be improved by chopping the top.

On a turret top vehicle (especially sedans), it is common for the back glass to be laid down. This often results in a hump above the back glass. To avoid this, it will be necessary to cut multiple slits in the top running forward and backward and then reshape the metal as required to eliminate the hump.

No one top chop method is a definite winner over another. It depends greatly on the style and purpose of the vehicle when it is finished. Raked back windshield posts will give more of a competition look, while stretching the top may provide better proportions. Ask different people with vehicles similar to yours how the top was chopped before you cut yours up. A stock height roof looks much better than a botched top chop.

Once you have determined how much to lower your hot rod's roofline and where to make the cuts, how do you actually go about doing the deed? The following steps are generic, but they should provide a pretty thorough checklist.

- Use masking tape to indicate the area to be removed. This will provide more of a graphic indication of what is going to happen before you cut. Since masking tape comes in a variety of widths, two lines of masking tape of the required sizes butted up next to each other will usually work well to indicate the material to be removed.
- Cut the top off and set it aside, out of the way.
- Cut the "chop" material out of the door and window posts that are still on the vehicle. This is easier to do than chasing the top around the shop floor.
- Reattach the top by tack welding it in place.

- Tack weld the doors closed, and then tack weld the door tops to the door bottoms. Fit the top of the door to the door opening and make adjustments for gap as necessary in the area that was cut.
- Finesse the body and top to fit together as perfectly as possible and finish weld.
- Finish welding the doors.
- Cut the garnish moldings to fit the new window opening. Secure both the top and bottom of the garnish moldings into their positions and tack weld them together to make sure they are aligned properly prior to welding them together.
- Finish bodywork as required prior to painting.

In their stock configuration, most vintage automobiles have square door corners, because they are easier to mass produce. An old custom trick is to round these corners.

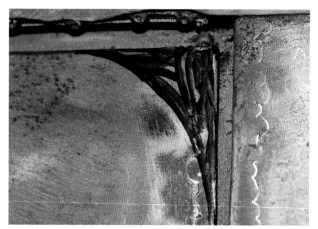

Begin the process by using some sort of template to mark the new rounded door outline. To better indicate the finished look, the corner that will be filled in is blackened with a black marker.

This is the passenger side of the same vehicle, but the door has already been rounded. The doorsill has not yet been filled to match.

A triangular-shaped piece of sheet metal, along with a tab that is blended into the body sill, has been welded in place to fill the once square corner.

With the filler piece welded in place, scuff it up to prep for a skim coat of body filler.

## Rounding Door Corners and Modifying Doorsills

Many hot rod bodies feature square-cornered doors in their stock configuration. While this may not be a big deal to some, the square doors on a round body is an aesthetic nightmare to others. The process of rounding door corners is simple in design, but tedious in execution, so it is not for the faint of heart. Perhaps not as much of a task as chopping a top, rounding door corners still requires a tenacious effort in fitting pieces accurately, welding them in place, and ultimately prepping the reworked area for paint. The actual process will vary somewhat depending on how the door is constructed, but the following instructions should provide sufficient guidance to complete the task, whether being performed on a door, deck lid, or hood.

Begin by determining an aesthetically appropriate radius for the new door corner. From most rounded door corners that I have seen, a radius of around 1 ½ inches seems about right. That is about the size of a small paint can, so you can

try tracing around one of those for a start. Hold the pattern (paint can, drafter's radius template, or whatever you decide to use) so that the radius becomes tangent with the two edges of the door and then mark around it. By using a black permanent marker, you can see the proposed cut line from a distance. Now step back and see if the line looks to have a pleasing shape and size. If the radius is too small, the work to be done will not be noticed. If the radius is too large, it will cause additional work to be done toward the inside of the door shell and simply will not look correct. If it does not look right, choose a different pattern now and redo it. Whenever you determine a suitable size template, mark all the doors, deck lid, hood, etc. that are going to be rounded so that they are consistent. If you're having trouble deciding what looks best, you can mark different choices on either side of the car so you can compare back and forth. Once you decide, mark the doors with the same curve.

Now use a plasma arc cutter or reciprocating air saw to cut the corner off the door. In theory, you would use this piece that you just cut off to fill the corner in the opening of the doorsill, but it may be just as easy to cut a new piece from new sheet metal to obtain a better fit. After cutting off the door corner, weld up any seam in the door that may have split. If the inside of the door shell encroaches upon this new door outline, trim the door shell to the correct shape and then weld in small patches as required.

Using the original corner piece that was removed from the door or new sheet metal, weld sheet metal into the corner of the doorsill to fill the area that was formerly covered by the corner of the door. It may be necessary to cut small strips of sheet metal to use as boxing strips. Then weld these in place to close off any open gaps.

Many of the vehicles originally manufactured prior to 1948 did not have doorsills like those found on contemporary vehicles. They simply had the edge of the floor exposed when the door was open. When the door was closed, the edge was covered, but the gap between the two was certainly not airtight. Creating a doorsill that will improve looks, as well as minimize road noise and wind noise, is a task that could be combined with rounding door corners. As seen in accompanying photos, this is accomplished by welding strips of sheet metal together to effectively box the area beneath the door and modify the bottom of the door to match. Getting the work right involves making patterns, taping them in place, and planning ahead to make sure everything fits before cutting any metal.

The edge of the floor is exposed in all of its unappealing stock originality. But that will all be changed by some relatively simple sheetmetal work.

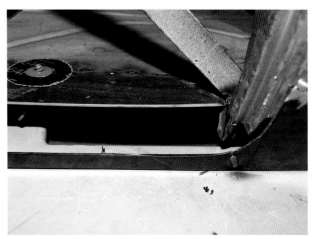

Sheet metal is being welded to enclose the area beneath the edge of the floor, between the front and back of the door opening. This will create a more contemporary-type doorsill.

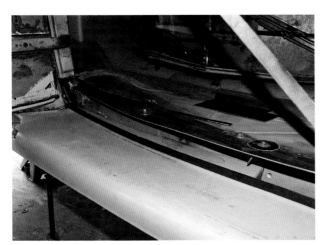

A wider shot of the sheetmetal filler plate that is being added to enclose the area between the floor and running board. This is one of those subtle improvements that may go unnoticed by the casual observer, but will greatly improve the looks of your hot rod.

With the doorsill closer to being finished, it is becoming less obvious. Subtle modifications that are tastefully executed are improving the pedigree (and value) of our hot rods.

By making patterns, cutting sheet metal, and then welding it in place, the door opening can have a more contemporary look. When finished, this will look much nicer than the original. Note that the bottom door corner has been rounded, just as the top corner has been.

Having a doorsill to work with allows more opportunity for utilizing rubber weather stripping to keep out moisture and noise.

Just as the door corners are rounded and the doorsill made to look more like those contemporary production automobiles, the vertical portion of the door opening gets some rework as well to blend it all in. Old-time hot rodders may not have cared about fit and finish or eliminating wind and road noise, but rodders of today expect the best.

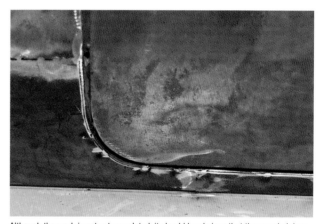

Although the work is not yet completed, it should be obvious that the rounded door corner and doorsill treatment will look considerably more "finished" than the original.

At this point, about all that is left to do on this modification is too apply a skim coat of body filler, and then block sand. Then it can be prepped for paint.

When the exterior door corner is rounded, it may require that the inner door shell be modified as well. Depending on the door being modified, it may just require that the inner shell be trimmed slightly and then closed off with a boxing plate on the bottom. Or, it may require rounding the inner corner somewhat, too. It doesn't matter; if you can weld, you can easily add whatever sheet metal is required to get it done.

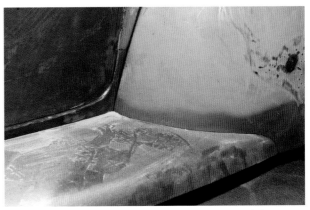

Not only were the door corners rounded on the rear side of the door, the lower corner was rounded on the front also. Originally, this lower door corner would have been very square and would overlap the edge of the floor, extending down to the top of the running board.

## Fenders and Running Boards

Apart from eliminating the fenders completely, most hot rod fender modifications have been somewhat limited to changing the headlights and taillights. Some rodders have widened the fenders, and others have modified the shape of the wheel well. Running board modification are typically even more limited, except for removing the rubber material and smoothing the running boards. Perhaps modifications to fenders and running boards simply do not catch our eye, or maybe modifications have been made so tastefully that we simply thought that is the way they were supposed to be.

In many stock configurations, front and rear fenders hang down off the body, with the running board connecting to the two in the front and back. This tends to make the running boards seem like an afterthought. Also running boards are often narrower than the fenders, which disrupts the otherwise smooth lines of the vehicle in question.

While many rodders and restorers alike feel that a '40 Ford is as close to perfection as has ever rolled off the assembly line in Detroit, there may be room for a bit of improvement. Nothing radical mind you, but something very subtle, that will go unnoticed by many. A modification that is being done to a '40 Ford Tudor sedan at Morfab Customs will integrate the fenders and running boards to improve the flow of the lower portion of the body's lines. Instead of simply spanning the gap between the front and rear fenders, the running boards are being reworked to look as though they pierce the fenders. Again, this is admittedly subtle, but sometimes the best modifications are just that.

This relatively stock '40 Ford coupe provides a look at the running boards and fenders in their stock configuration. Although the rubber-covered running boards are practical, they just are not stylish enough for some rodders. To some folks, the fenders come to an abrupt stop, with the running boards simply bridging the gap between the front and rear. Still, this '40 is no slouch.

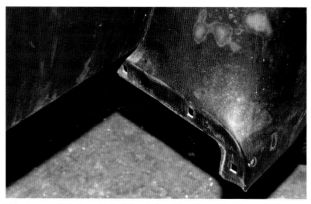

This close-up shot gives a look at the stock configuration of the left rear fender of the '40 Ford Tudor seen in many of these photos. The square holes are for the mounting bolts to secure the back end of the running board to the fender.

Now on the opposite side of the car, we see how the rear fender has been reshaped. The lower portion of the fender has been removed from the fender and grafted onto the rear of the running board. The overall shape of the assembly is basically the same, but the seam is now horizontal, rather than vertical, and therefore flows better with the other lines of the body.

Weld-through primer has been applied to prevent any rust from forming on the otherwise bare metal. This keeps the sheet metal clean but can be welded through without the need to grind or sand it off, as would be required with other types of primer.

With a little bit of hammer work, a bit of filler, and some primer, the fender portion of this modification is nearing completion.

On the back side of the fender, a piece of sheet metal is bent into an angle and then curved to conform to the curve of the fender by using a shrinker/stretcher. This flange was then welded onto the back side of the fender to serve as a mounting point for the running board.

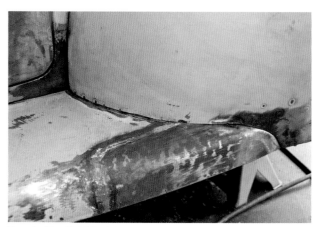

Those familiar with '40 Fords know that the front portion of the stock running board curves in slightly, with its edge intersecting the front fender approximately half way between the body and the edge of the wheel opening. To smooth up and straighten the lines, the front end of running board was widened to be more in line with the outer edge of the fender.

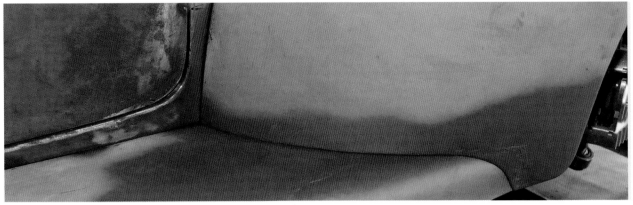

With the cutting and welding done and some weld-through primer applied, this area of the car looks much smoother already. It should look great when it has all been sanded, primed, and painted.

Much like on the rear fender, a sheetmetal flange was fabricated and then welded to the inside of the fender. The running board can then be secured to the fender from underneath.

## Straightening

During the course of building a hot rod, you will have many opportunities to straighten metal. The straightening may be required to repair obvious collision damage or it may be in order to align panels, whether they are original or replacements. Rather than just whaling away with a hammer in an attempt to straighten the metal, use a hammer and dolly, along with some finesse to do it correctly.

### Hammer and Dolly

The correct way to remove a dent is to essentially "undo" the damage. That is, hammer the dent out in the opposite progression of the way that it happened. On a simple dent, such as a baseball hitting the middle of a panel, this is a simple task to undo. In a collision, however, reading the dent is usually more involved.

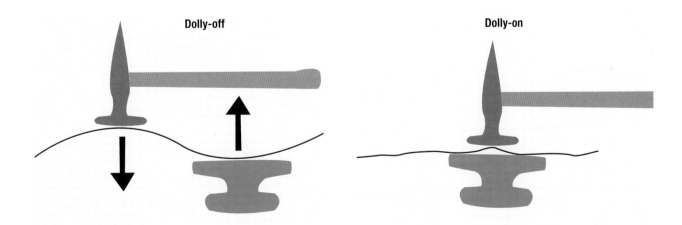

Most, if not all, sheetmetal hammer work should be done accompanied by a dolly behind the sheet metal. Dolly-off hammering is used for initial sheetmetal straightening, while dolly-on is used for flattening the metal after straightening it.

107

As they came from the factory, rear fenders on '40 Fords were relatively oval when viewed from the top. When viewed in profile, the extreme rear starts with a nice curve that seems to flatten out. It may be that they were designed that way or that manufacturing processes of the day limited the amount of compound curvature that could be gained from the sheetmetal press.

What would seem natural is for the fender to continue its natural profile by extending to the lower edge of the fender with the same radius so that it doesn't look flat at the end. Mike at Morfab Customs is extending the rear fenders by welding on a sheetmetal extension that could have been made by hammer forming a piece of sheet metal over a buck or possibly on an English wheel.

If you look close, you can still see approximately where the new piece of sheet metal is welded to the original fender. With the extension, the fender retains its natural curve through a longer distance, with a short compound curve tying the outer and inner portions of the fender together more gracefully.

In raw steel, we can gain a better perspective of how the fender looks with its extended profile. The masking tape is protection for the new taillight.

Pulling off the masking tape, we see that the stock "Chevron" taillights have been replaced by the teardrop-style taillight from the '38-'39 Fords.

When two vehicles collide, or a tree or utility pole jumps out in front of your car, dents become much more complex. During the initial impact, sheet metal is pushed inward. Since it is a formed piece that has been stamped into a predetermined shape such as a fender, it already has some forces built in. When these forces are combined with the force of the impact, the sheet metal surrounding the initial point of contact typically bulges outward as a reactionary force. In this simple example, the process would be to hammer the bulging sheet metal back into place and then hammer the dented sheet metal back out from the inside or pull it back out from the outside.

In most situations, using a hammer by itself (without a dolly behind the metal) will cause a larger-than-desired area of metal to move inward. This is usually not what is needed to repair a dent in a piece of sheet metal. By using a dolly behind the panel and a hammer in front of the panel, the sheet metal can be worked in a more predictable manner as the dolly focuses the force of the hammer.

Dollies are used in two basic ways: dolly on, or dolly off. When hammering on the dolly, the dolly is located behind the sheet metal and directly beneath the hammer blow. This method is used to knock down high spots or to smooth ripples within the relatively small size of the dolly. Hammering off the dolly is done by hitting the surface of the panel adjacent to the dolly, rather than on it. This causes the dolly to push outward while the hammer pushes inward and is typically used on larger areas of repair.

## Bending

Sheet metal can be bent in a variety of ways, depending on whether you desire a crisp, uniform bend; a uniform nonsquare bend; or if you have need for a freeform shape. Each type requires a different approach.

For uniform bends, such as for making a square fuel tank, a sheetmetal brake or box and pan brake is the way to go. You should make a pattern first so that you can determine where to actually make the bends in the metal. Additionally, you must consider that the sheet metal used will have a particular bend radius, which effectively shortens the metal somewhat.

If you are making a dash panel, hood side, or other sheetmetal component that will have a bare edge, that edge can be made to look substantially better by hammer forming the edge of the metal over the rounded edge of a piece of hardwood or construction-grade particle board used for making kitchen countertops. Refer to "Hammer Forming" later in this chapter for more details on how to do this.

For creating freeform shapes, a beater bag and a hammer are pretty cheap while an English wheel is expensive. Both require plenty of practice and the ability to read the metal in order to make duplicate pieces. For more information on using an English wheel, refer to the "Curving" section later in this chapter. *Continued on page 113*

# PROJECT 8
# Building a Fuel Tank

You really have to be careful when you mention custom fabricated fuel tanks. Due in part to liability and other insurance concerns, many people who may be perfectly capable of fabricating a gas tank simply will not consider doing so, or they will charge you a fortune for their services. While the process must not be taken lightly due to the potential safety concerns, building a custom gas tank for your hot rod is a relatively simple process. However, all welding must be done by a competent welder and the tank tested for leaks prior to any gasoline (or other fuel) being dispensed into the tank.

The first step is to determine the desired shape and size of the tank, along with the method of securing it to the vehicle. Once you determine a method of securing the tank that does not require any holes in the tank itself, the next step is to draw the patterns for the sheet metal. At the risk of oversimplifying the task, at this point your goal is to construct a closed box. As long as you can create the pattern and bend the sheet metal to match the pattern, the proposed fuel tank can be virtually any shape that you desire it to be (or that available space permits it to be). Just remember that for a gas gauge to work properly, the gauge float will need ample room to swing and the shape should be uniform.

For a rectangular gas tank, only two pieces of metal are required for the actual tank. One is bent into a "U" shape to create the front, bottom, and back of the tank. Another is bent to create the top and both ends. Flat plates to be used as baffles to dampen the momentum of fuel sloshing back and forth during cornering should be welded in place as desired as the tank is being welded together. They should be designed so that they do not trap gasoline in any one part of the tank and do not interfere with movement of the gas gauge float. By using the formula of 1 gallon = 231 cubic inches, you can determine how much fuel your tank will hold.

Before the tank is welded together, the filler neck should be installed and a hole drilled for the fuel level sender. There are a couple of things to remember when constructing a gas tank. For one, unless you use self-tapping screws for securing the fuel level sender to the tank, you will need to locate it close enough to the fuel filler that you can reach in with a wrench to tighten the mounting bolts and nuts. If you already have the gauges and therefore the instructions for adjusting the fuel level float, you should make the adjustment as necessary before welding the internal baffles in place. For those who may not be familiar with fuel level floats, the length of the float arm is adjusted according to the depth of the tank, so the baffles should be located so that they do not interfere with proper operation of the float. Additionally, a bung should be installed so that the fuel line can be connected.

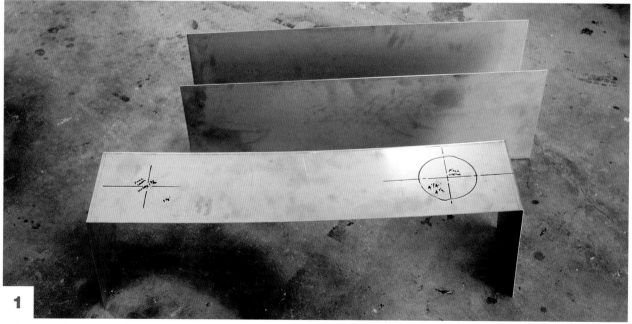

**1**

These are the major components of the gas tank: one piece of stainless steel that forms the front, bottom, and back and another piece that forms the top and both ends.

**2**

To make drilling the holes for the fill valve and the fuel gauge sender easier, they were done before the tank was welded together. In this photo, two lines are marked on the inside of the top panel. The intersection of the lines will be the center of the hole for the fill valve.

**3**

The key to drilling holes in stainless steel is using a slow drill speed and keeping the drill bit cool by keeping it lubricated with WD-40 or other similar lubricant.

**4**

Since I was using a hole saw that had a pilot bit that protruded longer than the full diameter teeth, the initial drilling was done with the sheet metal sitting on a piece of scrap wood siding. As I got closer to drilling through, the metal was supported by two scrap blocks of wood so that I could drill completely through the sheet metal.

**5**

After two holes were drilled using the same procedure, but different size hole saws, this is what the unwelded gas tank looks like. Other than the welding (which someone else will do), the hard part is done.

**6**

The fill valve and the fuel level sender were used as templates to mark the necessary holes for mounting them in the top of the tank. These smaller holes were made with a drill bit, rather than a hole saw.

**7**

Mounting studs welded to the flexible ring/gasket assembly hold the fill valve. The mounting studs extend up through the top of the gas tank, and the fill valve is then secured with stop nuts on each stud.

**8**

Apart from welding, the tank is essentially complete at this point, showing that this is an easy project to complete if you are in need of a gas tank. Since it will be a gas tank that must not leak, it must be assembled by a skillful welder. Since it has not yet had gasoline in it, you should be able to find someone capable of performing the welding tasks if you are not comfortable doing it yourself.

**9**

This is the fuel level sender and float assembly, a component that is usually included with aftermarket gauges when they are purchased in kits. These senders usually include instructions on how to adjust the float, depending on the height of the tank.

**10**

After being welded together, this is what the stainless steel gas tank looks like. Being stainless, it could be polished to a mirror-like finish if desired. The nipple at the right side of the photo is for the fuel line connection while at the opposite corner is a vent. Prior to putting any fuel in the tank, it should be tested for leaks with water and then thoroughly dried.

*Continued from page 109*

## Curving

Curving metal, as for a hood top or a fender, can be done in at least a couple of different methods. If you have ever sat through a boring presentation where a paper program was handed out previously, you can relate to how metal can be curved. As you no doubt subconsciously rolled that paper program in your hands, metal can be curved in much the same way. As long as the sheet metal or aluminum is thin enough to be curved, it can be (and has been) pushed around a welding gas tank or piece of PVC pipe to form a curved piece. An important thing to remember is not to exceed the arc the sheet metal can tolerate before bending or to curve it more than you want. Work the metal slowly and multiple times to achieve the correct arc, rather than rolling it too much and then trying to flatten it. Refer to "Making a Three Piece Hood" for more on this process.

### English Wheel

Another method of curving metal is to use an English wheel (a.k.a. wheeling machine). This method has been around for as long as the automobile and is often associated with custom coach building. Before automobile manufacturing facilities began using large, hydraulic presses to stamp fenders and other curved sheetmetal components, those panels were rolled on an English wheel. On extremely large panels, smaller wheeled shapes were made and then welded together.

To use an English wheel, a piece of aluminum or mild steel sheet metal is moved back and forth between the flat, upper wheel and the curved, lower wheel (or anvil). As the sheet metal is passed back and forth, it becomes thinner, which allows the metal to curve over the anvil. The larger the radius of the anvil, the lower the crown will be in the curved metal. The smaller the radius of the anvil, the higher the crown will be. You should begin with a lower crowned anvil and gradually work up to a higher crowned anvil as the work progresses. Starting with a crown that is too high will mar the sheet metal. Likewise, begin with just a slight amount of pressure on the sheet metal. The metal should not be able to slip through the wheels. If the pressure is correct, the first few passes will yield some shaping, but they will not mar the metal.

As the sheet metal becomes thinner (with each pass between the upper and lower wheels), the pressure between the wheels will decrease. However, the pressure between the wheels is easily adjustable by an adjustment knob located near the anvil or by a kick wheel located near the floor. For many, the lower kick wheel is preferred as it allows them to keep both hands on the metal being wheeled.

Repeated passes that are parallel with each other and that slightly overlap, along with correct pressure, and the use of the appropriate crown anvil are the keys to achieving the desired results when using an English wheel. If too much pressure is used, the sheet metal will be marred and uneven, while too little pressure will take too much time. Moving the sheet metal diagonally between the upper and lower wheels will cause the surface to look wavy. Since this process is stretching the sheet metal, the passes must overlap slightly so that the metal stretches evenly.

Quick tips for using an English wheel:

- Avoid starting with too much pressure on the metal, as this will result in long bumps rather than a curve in the metal.
- Gradually increase the pressure, but change the pressure only between passes.
- Work gradually to a higher crown to avoid overstretching the metal.
- To create a compound curve, make short passes on the area that should have more curve, and then overlap these first passes while wheeling the entire part.
- Avoid running the edge of the metal through the wheel, as this will stretch the edge and potentially cause fitment problems.

Starting with a flat piece of sheet metal and an English wheel, you can make curved panels. This is useful for making replacement door skins, cowl panels, or a multitude of custom pieces.

Pressure between the upper and lower anvils is adjusted by the kick wheel. This allows the operator to keep both hands on the sheet metal being worked, while making adjustments on the fly.

The tracking marks also help the operator duplicate the process when making similar and often opposite panels.

Rather than starting with a high crowned anvil from the beginning, it is better to start with a lower crowned anvil and then increase the crown later in the process. If too high a crown is used too soon in the process, the metal will be stretched too much.

Setting the sheet metal on a flat surface shows just how much curve has been wheeled into the metal with just a few quick passes with a relatively low crowned anvil. From looking at the narrow edge, it is easy to see that more passes have been made near the middle, rather than at the flatter edge.

Worked on a higher crown anvil, the once-flat sheet metal now begins taking a curved shape more progressively. As Kris wheels closer to the edge, a more pronounced shape begins to develop.

Each pass between the upper wheel and the lower anvil creates a tracking line on the sheet metal. Keeping the lines parallel helps the operator know that the sheet metal is being kept square with the wheel.

When compared to a previous photo, it is obvious that the higher crown anvil creates more curve in the sheet metal.

## Rolling

A hood top (the center piece of a three-piece hood) for a '32 Ford, as an example, has a relatively flat curve in the middle and then has a curve of much tighter radius toward each of the sides. Since the front and back edges of this hood are parallel, it is fairly simple to make this type of hood by rolling the metal around a cylinder, such as a welding gas tank. The cylinder must be longer than the metal being rolled around it and should be smooth on the outside to avoid putting dents or wrinkles into the sheet metal.

You should first determine a centerline between the cowl and the grille shell. As a point of reference, you can use a piece of masking tape stretched between the cowl and the grille shell. Then make a line on the sheet metal to represent the centerline. Measure the distance between the lip on the cowl and lip on the grille shell where the hood will fit and transfer this same measurement to the centerline on the sheet metal. Mark lines that are perpendicular to the centerline at the beginning and end that represent the front and back of the hood. Do not trim it yet, however. Measure along the cowl from the centerline to where the flat curve of the center compounds into the smaller radius curve. Transfer this measurement to the left and right of the center on the line that represents the back of the hood. Now repeat this process on the grille shell for the front of the hood. Now draw a line between the marks on the line representing the back of the hood and the respective sides for the front of the hood. Since the grille shell is narrower than the cowl, these lines should be closer at the front and farther apart at the back. Now measure from the centerline at the back of the hood to the point on the cowl where the side edge of the hood top would be. Transfer this measurement to the sheet metal along the line that represents the back of the hood. Do the same along the grille shell and transfer those measurements to the line that represents the front of the hood. To be on the safe side, allow a couple of inches on all sides and cut out the sheet metal.

Now place the sheet metal top side down on a worktable. Place the welding tank or other cylinder on the centerline of the sheet metal. Gently and smoothly lift up on one side of the sheet metal so that it begins to curve around the cylindrical form. Now lift up on the other side the same amount. Lay the new hood on the vehicle to check if the curve is about right. If it is not, continue the process, checking often to make sure you do not bend the metal too much. When the center portion of the hood is curved the correct amount, move the cylinder so that it aligns with the line that represents the point of compound curve where the curve changes to a smaller radius. Lift up on the outer edge of the metal until it is the correct shape to match the cowl and grille shell. Check often and then repeat as necessary until the correct shape is obtained. Then do the opposite the same way. When the sheet metal is curved as required, trim the new hood to fit.

*Continued on page 124*

# PROJECT 9
# Making a Three-Piece Hood

**A** very common hot rod modification is to discard the stock four-piece hood and replace it with a three-piece unit. The three-piece hood allows the owner certain variations, as the hood top can usually be removed and the hood sides left in place or the sides removed and the hood top left in place. Of course, you can also use all three pieces or leave them all off as desired. Keith Moritz and Ralph Stogsdill were making a new hood for their M&S Racing '34 Ford Bonneville record holder and allowed me to follow the process.

**1**

The crewmembers at M&S Racing decided that a new aluminum hood would be a good addition to their Bonneville record car. While it has not been determined if the hood actually made the difference, Ralph and the boys did break the record for their class.

**2**

This profile shot of the engine compartment shows what the front, back, and lower edges of the hood must conform to. Since this hood is a complex shape, a pattern will be required.

**3**

Keith Moritz starts by using masking tape to tape several pieces of poster board together. Poster board is a suitable material as it is flexible, yet will hold its shape rather than collapsing under its own weight.

**4**

Remember to have sufficient pattern material to cover the entire area that must be covered. Excess can be trimmed away later.

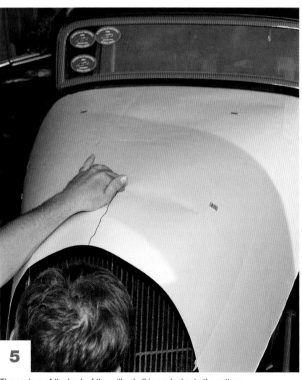

**5**

The contour of the back of the grille shell is marked onto the pattern . . .

**6**

. . . and then the excess poster board cut away from the front portion of the pattern.

**7**

A metal straight edge is used to mark the centerline of the cowl to provide a point of reference.

**8**

The narrow gap between the two pieces of masking tape represents the centerline of the cowl. This allows for a more precise location than inconsistently choosing one side or the other of a marked line.

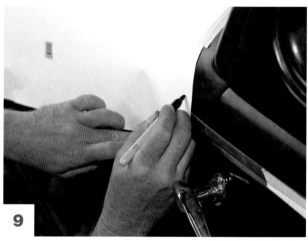

**9**

Keith uses a permanent marker to indicate body line reference points on the pattern.

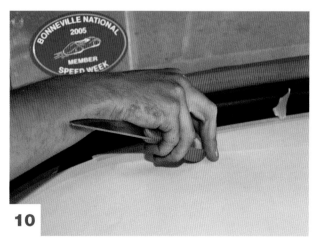

**10**

By rubbing a hard surface such as a scissors handle along the body line of the cowl, the line is transferred to the pattern material.

**11**

Before any cutting is done, Kris heavies up the creased line just made by tracing it with a permanent marker.

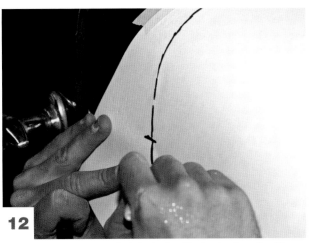

**12**

The heavy vertical line represents the back edge of the hood on the pattern. The short horizontal line represents a body line that will serve as a reference point.

**13**

After the outline of the hood is marked on the pattern, the excess is cut from the front and back.

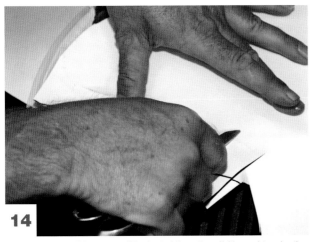

**14**

After cutting most of the excess off the front of the pattern, Keith can determine the actual line to be cut. He rubs the scissors over the poster board at the back edge of the grille shell, leaving a slight crease at that line.

**15**

If the crease is obvious enough, it can be cut as is. If necessary, trace over the line with a marker to make it easier to follow while cutting.

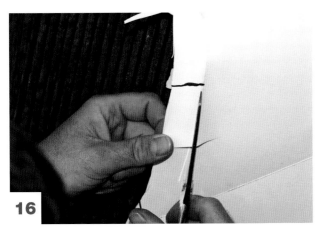

**16**

Keith continues cutting around the grille shell. Right at about Keith's left thumb is a reference mark that will be used to match the pattern with reference points on the grille shell.

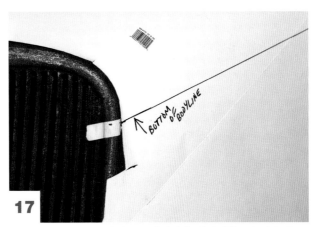

**17**

The line shown represents the bottom of a body line that will be extended onto the hood.

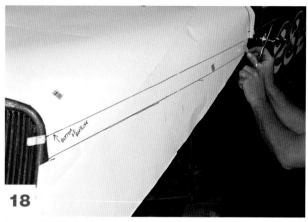

**18**

Approximately 2 inches below the bottom of the body line shown in the previous photo is a line that represents the extreme edge of the hood top.

**19**

Excess pattern material can now be cut off and discarded.

**20**

Keith now begins removing masking tape that was used to hold the large pattern in place.

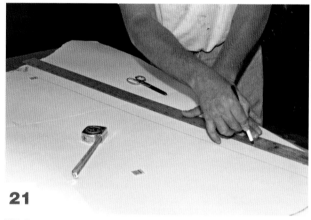

**21**

With the pattern removed from the car and lying flat on a table, a straight edge is used to connect reference marks that represent the body line and edge of the hood.

**22**

After the perimeter and all of the reference marks and lines are made on the pattern, the pattern can be cut out.

**23**

The pattern is now laid out on the aluminum sheet so that the grain of the aluminum is running the same direction as the bends that will be made. If the grain is running against the bend, the aluminum will be more likely to distort. Leaving enough room for final trimming later, the portion of the aluminum sheet to be used for the hood is cut down to approximate size for ease of handling.

**24** The poster board pattern is then secured to the aluminum sheet with masking tape.

**25** Another hood was available to verify the approximate location of where the bend should begin. If an existing hood is not available, you can determine this location from the grille shell and the cowl.

**26** Keith and Kris use a straight edge to connect the front and rear points that define the approximate line where the hood will need to be bent. Do this on both (driver and passenger) sides of the hood.

**27** The perimeter of the pattern now must be transferred to the aluminum sheet. Kris does this by simply tracing around the pattern with a permanent marker. Any reference lines, body lines, or other critical marks must be transferred as well.

**28** With the pattern out of the way, lines are drawn on the aluminum to connect the various pairs of reference points.

**29** Keith normally uses a gas welding tank as a form to wrap the aluminum around, but this hood is longer than his tank, so a piece of PVC will be used. To act as a stop to keep the pipe from moving, a piece of square steel tubing is clamped to the table top with a pair of C-clamps.

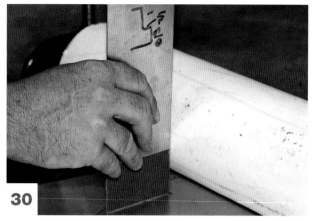

**30**

A square, or in this case a piece of sheet aluminum that is cut square, is used to align the edge of the PVC pipe with the line that represents where the bend should begin.

**31**

The lower edge of the aluminum sheet is then pulled upward around the edge of the PVC pipe, forming a nice smooth curve in the process. To help prevent kinking the aluminum sheet, a steel tube that is longer than the hood is used to pull the aluminum up evenly.

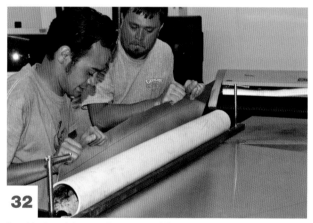

**32**

Do not be too quick to cut off excess material, as you will need something to hang onto when bending the aluminum. The process is repeated on the opposite side of the hood.

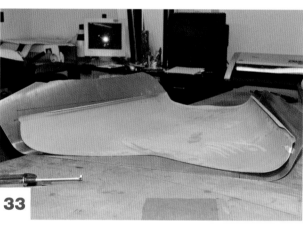

**33**

With a "store bought" hood top used for comparison, it is easy to see that the new hood is fairly close to being the correct shape. A larger-radius PVC pipe (such as the welding gas tank that Keith normally uses) would have duplicated the bend a little more precisely.

**34**

Much of the excess is cut off with a shear. Final trimming will be done by hand.

**35**

The outline of the actual hood is difficult to see, but it is there. These curved shapes can be cut with a plasma arc cutter, a hand shear, air saw, or aviation snips.

**36**

With the hood set in place on the coupe, it is checked for proper fit, adjustment, and final forming around the edges.

**37**

A bit of hammer and dolly work is required to fit the new hood to the vintage grille.

**38**

After cutting out the hole in the hood for the scoop, the hood was again checked for proper fit. All that is left to do now is to use a set of custom-made dies to roll the body bead into the sides of the hood and to install the Dzus fasteners.

**39**

Tested by a Bonneville run that set a new class (XF/VFALT) record (130-plus miles per hour), the new hood apparently fit just fine.

*Continued from page 115*

## Punching Louvers

While louvers were originally stamped in the sheet metal of early hot rods to allow for more airflow around the engine, with today's improved fans and radiators, they are as much for form as function. Louvers are available in a wide variety of shapes and sizes. The size of the louver is up to you, but louvers commonly found on hot rods are typically from 2 ½ to 4 inches wide.

Prior to punching louvers, strip the sheet metal of paint and primer on both sides. Punching a painted part will not allow the louvers to be as crisp as they should be. Plus, the process of punching louvers will no doubt chip or scratch some of the paint, so a repaint would be necessary anyway. Now mark the layout of the louvers on the inside of the panel. For the best results, rows of louvers are usually laid out parallel to each other. You should first determine if there will be an even or an odd number of rows of louvers. If there is to be an odd number, the centerline of the panel

will also be the centerline for the middle row of louvers. For an even number of rows, you will need to offset from the centerline. To determine the location for the next centerline, measure the width of one louver plus the spacing that you desire to have between rows. This is the same as half of one louver width plus the spacing plus half of the next row's louver. With the centerlines of the rows laid out, begin marking the actual location of the louvers. Regardless of the shape of the louver, the back side will be straight. Mark this location along the centerline for each louver to be punched. If you punch louvers on a regular basis, you will undoubtedly create some sort of notched stick or template to use for layout, rather than measuring each and every louver location.

Once all of the louvers are laid out, the louver punching can begin. As the metal being punched must be moved into position for each louver, it may be necessary to have an assistant if the panel is very large or otherwise awkward.

Louvers can be functional or decorative, large or small. The small louvers on the tailpan of this World War II bomber-inspired Model A help to enhance the aero theme.

The large louvers are the same as the stock configuration, although the hood material on this hot rod is aluminum. Original style latches and handle indicate that this may actually simply be a replacement, stock-style hood.

The louvered deck lid and belly pan on this five-window coupe make it look like a true dry lakes racer. Notice that the edge of each louver is pinstriped in red for that extra bit of detailing.

On the underside of the hood of my track T, the layout lines of the louvers can still be seen. The layout lines represent the centerline of the louvers and should be spaced as necessary to allow for the desired spacing between the finished louvers.

Vince Baker of KBS Fabricators uses an experienced eye and careful hands to position a hood side prior to pressing the foot pedal that operates the louver press.

This is the new aluminum hood top that KBS Fabricators made and louvered for my track T. It sure looks better than the fiberglass hood top that was part of the kit.

## Bead Rolling and Making Flanges

When using a bead roller, the thickness of the sheet metal being used will have a significant impact on your results. For aluminum, a thickness near 0.040 inch works well, while 20- or 22-gauge steel is appropriate. Using material that is thinner than this will typically result in panels that are distorted. Additionally, if using a flanging die, the upper and lower rollers must be offset by the thickness of the metal, or the panel will distort.

With a properly shaped die, an edge similar to the stock edge of this vintage dash can be rolled into a dash panel or other piece of otherwise flat sheet metal. For a dash in particular, some sort of rounded edge is more desirable than a simple straight edge.

## Hammer Forming

Hammer forming is commonly used to place a rolled edge on a piece of metal. Examples are the lower edge of a relatively flat dash or an opening in a hood for a blower scoop. The rolled edge simply looks smoother than a plain hole cut in a piece of sheet metal and also adds a bit of rigidity. For technique, a light touch and finesse are better than brute force.

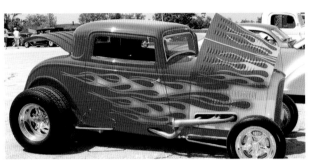

Rather than having a simple flat edge in the hood where the exhaust exits, this hood side has a hammer-formed edge. This provides an attractive edge that also provides some rigidity to the hood side.

You will need to create a hammer form base and a hammer form top that is the same size and shape as the metal you will be working with. For a one-time-use hammer form, you can use particle board, but hardwood such as oak should be used if the form will be used for other similar projects. If using particle board, put two pieces together with drywall screws to give the hammer form more rigidity. The edge of the opening in the hammer form base is what will define the rolled edge, so it needs to be uniform in shape and smooth. To uniformly round the edge, use a router and an appropriate shaped router bit. After shaping the opening, smooth it out with sandpaper. The final product can only be as good as the form, so take your time and get the hammer form base as good as you can.

Place the hammer form base on a workbench, and then place the metal atop the hammer form base so that the edge to be formed is aligned with the hammer form base. To prevent the metal from developing lots of ripples during the process, place the hammer form top on top of the previous pieces. Make sure that they are all properly aligned and then clamp all three pieces together. The metal being formed should be sandwiched between the hammer form base and hammer form top. Use as many clamps as you can muster to eliminate any movement between the three pieces.

Using a forming hammer, start tapping the metal downward closest to the hammer form top. Work your way around the entire edge, gradually moving downward and keeping the metal tight against the hammer form base in the process. You should avoid hitting the metal too far away from the hammer form base, or you will risk bending the metal, resulting in a crease. As it becomes necessary to move clamps, move them only one at a time to help keep the flat portion of the panel flat.

*Continued on page 132*

# PROJECT 10
# Building a Flat Dash with a Round Edge

For my track T roadster, I had tried my hand at making a steel dash. I was not pleased with my hammer-forming ability, so I asked Keith Moritz of Morfab Customs to make a dash for me. The dash could be made of steel, but for an open vehicle, aluminum would probably be a better choice of material. On an open car, the dash, as well as other internal components (upholstery, carpet, etc), are subject to the elements. If the dash is completely painted (front and back), it does not really matter. However, if a steel dash is not painted on the back side, or if it gets scratched, it will rust eventually, while an aluminum (or stainless steel) dash will not. Since Keith is an ace fabricator, the new dash will surely look good, regardless of what it is made of.

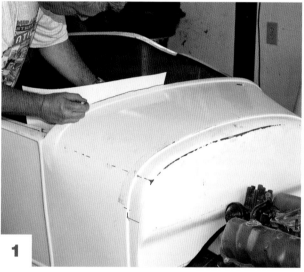

**1**

To get the curve of the top of the dash, a piece of poster board was aligned against the dash centerline. A line was then drawn along the top edge of the dash. The poster board was then flipped over to achieve a mirror image for the lower side of the dash. *Photo by Sandy Parks*

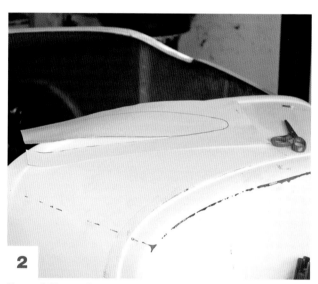

**2**

Since available poster board was not quite wide enough to fit the entire dash, a half pattern was made. The pattern was then cut out with a pair of scissors.

**3**

After drawing a centerline on a piece of pine lumber that is wide enough and long enough to fit the pattern on, the pattern is aligned with the centerline. Then the centerline is traced around, indicating where the wood needs to be cut.

**4**

I used a scrap piece of pine shelving because it was available, but for repeated use, oak or some other hardwood would be more suitable.

**5**

Now that the entire dash panel has been traced onto the wooden buck, the buck needs to be cut out. A jigsaw, scroll saw, or band saw should work for this.

**6**

After cutting out the wooden buck, check it for fit. If any more cuts need to be made, you might as well make them before smoothing the edge.

**7**

Keith uses a router and an appropriate bit to shape a smooth round edge along what will be the bottom of the wooden buck. This rounded edge is what the aluminum or sheet metal will be hammered against, so the edge must be smooth and uniform for the best results.

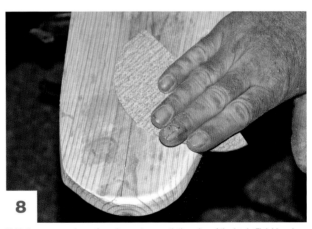

**8**

Keith then uses a piece of sandpaper to smooth the edge of the buck. Finishing does not have to be extensive—it just needs to knock off any rough edges.

**9**

It would be difficult to cut the excess material off of the dash after it is bent, so Keith uses a pair of dividers to trace a line a uniform distance from the lower edge of the buck.

**10**

Keith then trims the excess off the sheet metal by passing the line just marked through the band saw.

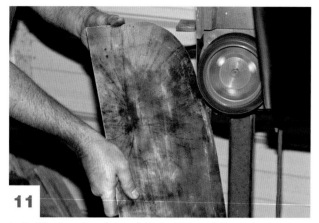

**11**

As should be done after cutting most sheetmetal pieces, Keith smoothes the edge on a belt sander.

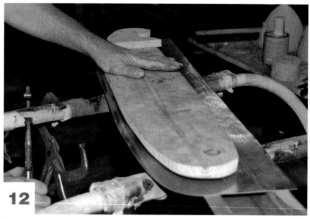

**12**

The next step is to position the buck correctly in relationship to the sheet metal or aluminum.

**13**

Then the buck and sheet metal must be clamped together, as there should be no movement between the two. Keith uses a piece of scrap sheetmetal hat channel and a couple of clamps to clamp the buck to the sheet metal.

**14**

With his trusty forming hammer, Keith begins hammering the lower edge of the sheet metal over the rounded edge of the wooden buck.

**15**

To prevent the sheet metal from distorting, the edge will be bent over in several passes, rather than all at one time. If you try to bend too much at one time, the sheet metal will "bottle cap" or crimp, which is not acceptable.

**16**

Having the sheet metal and buck clamped together securely, having patience, and using finesse rather than brute force are the key ingredients to success when hammer forming.

**17**

With the dash mounted in the car, the next step in laying out the gauges is marking a vertical centerline and a horizontal baseline. The intersection of these two lines will be the center of the speedometer, with two smaller gauges on either side.

**18**

I used a circle template to draw in the gauges. Once the spacing looked correct on one side, it was duplicated on the other side. Although the layout looks fine (at least to me), I should have raised the entire layout between ½ and ¾ inch to allow for some additional room for the ignition switch. Remember that even though the layout may look fine, there needs to be sufficient room on the back side for the wiring.

**19**

Good, bad, or indifferent, the first hole was drilled, so we were off and running. At least this type of dash would be easier to reproduce than, say, a dash for a '40 Ford.

**20**

Since everything went OK while drilling the large hole for the speedometer, I changed to a smaller hole saw and drilled the four holes for the smaller gauges. In the limited confines of the track T, all of the gauges will be readable by both the driver and passenger.

**21**

With the holes drilled, a grinding stone on a pneumatic die grinder was used to open the holes slightly to actually fit the gauges and to smooth the edges. The gauges were then installed and secured from the back side.

**22**

After realizing that I had forgotten to include the ignition switch in the layout, I determined that there was adequate room between the speedometer and the gauge to its left. The closest size drill bit was used and then the hole was opened slightly by using a cone-shaped stone on a pneumatic die grinder.

**23** Progress on getting the hole size correct could be checked from the outside. When the size was correct, the ignition switch was installed from behind and then secured with a bezel nut that threads onto the switch from the outside.

**24** Headlight switch, all five gauges, and the ignition switch all fit nicely into the dash. Three tiny lights (turn signal and high beam indicators) will be installed in the dash behind the steering wheel or centered over the speedometer. When finished, the dash will be painted the same color as the body.

*Continued from page 125*

## Fastening

Since we have discussed cutting and shaping metal, it would seem appropriate to discuss how to fasten these metal pieces together. Depending on the application of your metal masterpiece, it may need to be fastened permanently or may need to be removable.

### *Permanent*

To permanently fasten metal pieces together, there is no doubt that welding is the way to go. Since welding is such a versatile skill, choosing the type and technique of welding for your application could fill a book all its own. If you want to improve your welding skills, I would like to suggest Todd Bridigum's Motorbooks Workshop series book, *How to Weld.*

### Smoothing a Bumper (Hiding the Bolts)

One way to smooth up the looks of the front or rear of your hot rod is to hide the bumper bolts. Presuming that you can weld, you can complete this task in less than a weekend, start to finish. The first step is to mount the bumper in the desired location, making sure that it is properly positioned (level and side to side). Then tack weld the bumper bolts in place on the outside face of the bumper. After doing this to all of the bumper bolts, remove the bumper from the vehicle, slip a large flat washer over each bolt, and solidly weld the bolt and washer to the inside of the bumper. Adding this flat washer to each bolt will prevent the bolt from pulling through and leaving a square dent in the finished front surface of the bumper. When finished welding, grind the bolt heads off of the outside of the bumper. If plastic body filler is used to finish the front of the bumper, the bumper can be painted to match or contrast with the body color, or even with chrome paint. To have the bumper rechromed, low spots will need to be filled with lead or brass. Consult your favorite chrome plating shop for their recommendations for how to prep for chrome plating.

Bumpers on many production vehicles from the late 1940s through the 1970s were secured by bolts that passed through the bumper and bumper bracket, with a nut on the back side. Whether painted or chrome, these bolts detracted from the otherwise smooth look of the bumper. By welding the bolts on to the back of the bumper and filling the holes, the bumper can be made to look much better.

### *Removable*

For attaching sheet metal panels that must be removable, quarter-turn fasteners and speed nuts are two favorites among hot rodders. Quarter-turn fasteners (commonly referred to as Dzus fasteners) have been securing hoods and countless other panels onto hot rods and all sorts of race vehicles for several decades and at speeds well over 200 miles per hour. When mounted properly, these will hold tight and can then be removed quickly and easily with a quarter-turn of the fastener. Speed nuts allow the use of bolts to fasten multiple panels together where the panels are too thin to thread and accessibility to the back side for a nut that is limited or nonexistent. Quarter-turn fasteners are a bit of a specialty item, so you may be required to order them from a speed shop. Speed nuts are commonly used in production vehicles, so they can be found at most auto parts stores.

Exposed bolts on the hinges of this pickup tailgate give it that old time, beast of burden appearance. The bolts used to keep the tailgate closed continue the theme. A note of caution, however: use locknuts on the bolts or the nuts may vibrate themselves off the bolts, allowing the tailgate to fall down.

### *Quarter-Turn Fasteners*

Consisting of two individual parts, quarter-turn fasteners are easy to install and can usually secure three layers of panels, depending on the thickness of those panels. All of the panels that are to be secured must have a hole drilled at a common location, so panel alignment prior to installation of the fasteners is critical. The size of the hole will depend on the diameter of the fastener, so check the required size prior to drilling any holes.

With the holes drilled in the panels, a special spring designed for this type of fastener should be located on the inside of the innermost panel. Locate the spring so that it is centered on the back side of the hole and then secure it to the panel with a pop rivet through the curl on each end.

The second piece of the fastener may be made of plastic or steel and have an oval, flat, or butterfly head. Except for the butterfly head, which can be turned by hand, heads have a slot. Special tools are sold for the purpose of turning these types of fasteners, but many rodders simply use a screwdriver or a coin. A relatively new style uses a female Allen head that can easily be tightened or loosened with a common Allen head wrench and is less likely to slip and scratch the paint. The head may also be a standard one-piece or a self-ejecting style that includes a mounting plate and a spring between the two pieces. To use the standard type, simply push the shaft through the holes in the panels and align the slot in the shaft with the spring on the back of the inside panel. As the quarter-turn fastener is turned clockwise, the shaft catches on the spring, pulling the panels tight in the process. The self-ejecting type is used the same way, but the mounting plate is fastened to the outer panel with pop rivets through its two mounting holes.

To remove the panels, simply turn the head counterclockwise a quarter-turn. The self-ejecting style will pop outward due to the enclosed spring, but will remain with the outer panel since it is riveted in place. The standard style must be pulled out after being loosened, but then must be kept in some sort of container to avoid being lost.

Like louvers, quarter-turn fasteners provide a competition look, due to their original use. The simple fact is that they work very well, do not take up much room, and are relatively cheap when compared to most other fasteners and latches.

Three views of a typical quarter-turn fastener, showing the slotted head at the left and the slot in the shaft that goes around the inner spring, which is mounted on the inside of the inner panel, shown at the upper middle. The lower right is similar to a section view, showing the mounting plate that gets riveted to the outer panel, the shaft of the head, and the self-ejecting spring.

### Speed Nuts

If you do not have easy access to the back side of a panel to use a nut and the metal is too thin to thread, speed nuts can be used to attach two or more panels. Speed nuts are available for different thicknesses, different size bolts, various widths, and various depths. The depth of the speed nut determines how close to the edge of the panel the holes must be drilled. One side of the U-shaped piece of metal has a relatively large hole, while the opposite side has a smaller hole if designed to be used with sheet metal screws or a slightly different arrangement if designed to be used with a bolt. In either case, the screw or bolt will tighten when inserted through both sides of the speed nut.

After the panels are aligned and mounting holes drilled, a speed nut can be placed over the hole on the inside panel. Align the second panel and then insert a sheet metal screw or bolt and tighten it in place.

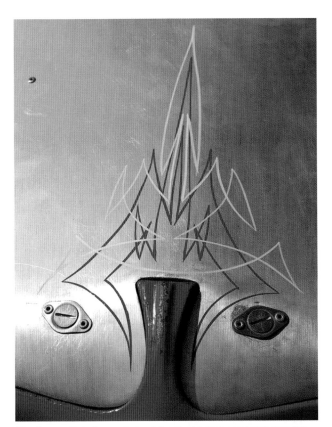

When in use, self-ejecting quarter-turn (a.k.a. Dzus) fasteners are held in place with two rivets through the mounting plate and the outer piece of sheet metal. A spring attached to the head and the mounting plate keeps the head in place but free to turn.

Speed nuts allow multiple panels to be bolted together in situations where a second wrench is not a feasible option, such as a transmission access panel. The speed nut slides over the edge of the stationary panel with the smaller hole to the side that cannot be accessed easily. Bolt holes in the stationary panel must be located close enough to the edge that the holes can be aligned. An appropriately sized bolt can then be used to secure the removable and stationary panels.

# Chapter 6
# Painting: Surface Preparation and Undercoats

Lots of "car guys" and "car gals" want to know how to paint their car. Since painting is such a popular topic, virtually every automobile magazine does at least one "paint special" issue each year. There are also several books on the subject available, with one of my favorites being MBI's *How to Paint Your Car* by David H. Jacobs and Dennis W. Parks. The one thing that some readers fail to realize is that the actual painting step is very simple and a very small portion of the overall success of a paint job.

The plain truth is that surface preparation is the key to a good paint job. No matter who actually squeezes the trigger on the spray gun, what brand or color the paint is, or what make or model vehicle it is going on, if the surface preparation is less than suitable, the paint job will be less than suitable as well.

## SURFACE PREPARATION
Regardless of the countless hours you may have spent modifying the body of your hot rod, you should take the time to make sure the rest of the body is as perfect as it can be prior to painting it. You may have a flawless top chop, sectioned hood, or countless other artful touches, but a minor dent, ding, or other blemish left unattended will detract from all of that. Now is the time to locate all of the little (or not so little) dents, dings, and scratches that did not seem important until now. After all, if you have made any of the typical or even not so typical hot rod modifications successfully, fixing these minor blemishes is relatively easy.

### Filling Low Spots
Most minor dings and other low spots can be eliminated, even if you do not have access to the back side of the panel. If you do have access to the back side of the panel, use a body hammer to gently work the ding back out so that a minimal amount of body filler is required. If the dent is less than about ⅛ inch deep, it can be eliminated by filling it with body filler. Begin by scuffing the paint in the area to be filled down to bare metal or epoxy primer. If you have multiple dents to fill, go ahead and remove the paint from all of the divots. Scoop an appropriate amount of body filler onto a mixing board. Don't mix more than you can apply before it will begin to set up. Apply the appropriate amount of hardener and use a plastic spreader to thoroughly mix the filler and hardener material

until it has consistent color throughout and no streaks. Use a plastic spreader to apply the filler to the dent, using a wiping motion in one direction across the dent. If done properly, the filler will stay in the dent and wipe off the area around it. Use this same method to fill the rest of the dents.

After the filler begins to cure, use some 50- or 80-grit sandpaper on a sanding block to knock off any high spots from around the dent. If you leave a gouge in the filler, it needs some additional time to cure. If sanding causes dust, you can continue sanding until the dent is sanded down to the correct level. If necessary, a second coat of filler may be applied to completely fill the area. After rough shaping with 50- or 80-grit sandpaper, use 140- or 180-grit sandpaper to blend into the surrounding area.

To get the cowl of the track T to match the contour of the windshield frame, body filler had to be added to the cowl. In order to get it right, I bolted the windshield frame in place (it attaches from under the cowl) and then worked the filler to match it. I found out later that I could have used strapping tape to cover the frame to keep the body filler from sticking to it.

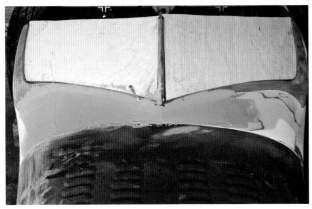

Since the windshield frame is made of three separate pieces, plywood "glass" is used to keep the three pieces aligned at the desired angle. The windshield frame will need to have the body filler sanded off of it when this process is completed. Although I have not done anything to finish the windshield frame for these photos, I'm leaning toward painting it to match the body.

On a small vehicle such as the track T, there is not much room to transition from one panel to another. So while working the cowl area, I kept the hood top in place to make sure the panels flowed correctly.

Extra attention has to be paid to the area adjacent to the hood so that the edge of the cowl is smooth and uniform.

Closer, but certainly not finished. Unless it just comes as a natural ability, bodywork will most likely test your patience during the course of a project. Even if it means not getting the project finished as soon as you'd like, don't proceed to the next step until each step in the process is done to the absolute best of your ability.

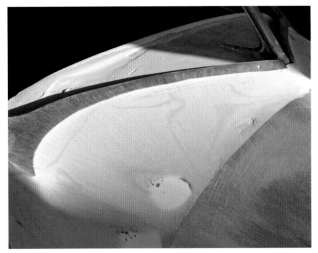

With most of the body filler sanded off the windshield frame, this is roughly how the cowl and windshield frame will look when finished. It was very tempting to leave the frame bolted in at this time and paint it in place, but doing that would allow the body filler to crack eventually.

## Sanding

While sanding too much is pretty much unheard of, sanding incorrectly is very common. It is not readily noticeable while the vehicle is in a dull coat of primer, but when a glossy coat of paint is applied, incorrect sanding will quickly become noticeable. As long as your techniques include (1) using a sanding block or board, (2) using the appropriate grit sandpaper, and (3) sanding in an "X" pattern, sanding is pretty easy. However, all too often the amateur body man gets lazy or tired, or tries to cut corners, only to minimize the positive effects of his hard work.

Many amateurs waste their sanding efforts simply because they do not use a sanding block or sanding board. Instead, they use the palm of their hand. For sanding to be done efficiently, the sandpaper must make full, even contact with the surface. If you squeeze the palm of your hand in several places, you will find that it is relatively soft in the

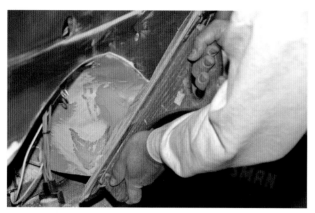

The exterior toe board area of the track T is flat in many places, so a long board is used to sand that area. In the rounded transmission area, it will be necessary to use a more flexible sanding block to get that area smooth.

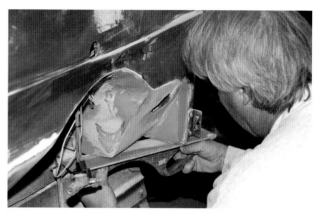

The exterior perimeter of the body and the underside of the floor are at the same level, so a long board is used to sand both of them as well.

palm, but harder at the knuckles and joints. Even though your hand can move the sandpaper across the body surface, more pressure will be applied at the knuckles and less pressure applied at the palm. This will cause waves in the panel due to the uneven pressure of the sandpaper as it crosses the body surface.

The appropriate grit sandpaper for the surface being smoothed depends on whether you are smoothing body filler, color sanding paint, or are somewhere in between. Each

step of the bodywork process has a range of sandpaper grits that are appropriate for that particular step. The actual grits that you use are relative to each other so you do have some latitude in the specific grit that you use for a task. Still, you must remain aware of where you are in the process whenever you pick up a piece of sandpaper.

Another amateur mistake is to sand in a simple back and forth motion, creating a flat spot or gouge in the process. The correct method is to sand in an "X" pattern. You can do this

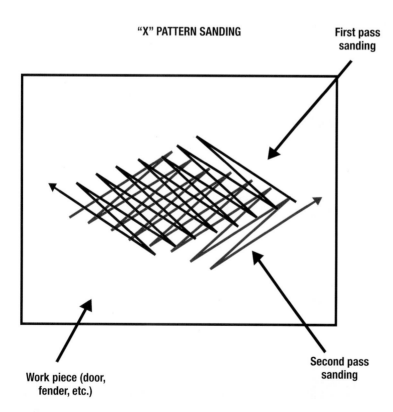

**"X" PATTERN SANDING**

**First pass sanding**

If you simply sand back and forth in the same place without moving laterally simultaneously, you will sand a gouge into the surface. As this sketch shows, you must move the sanding board or block forward and backward while moving to one side or another to get the area smooth. To get the area as flat as possible, sand in opposite directions as shown in the second sanding pass.

**Work piece (door, fender, etc.)**

**Second pass sanding**

by pushing the sanding block sideways as you make several diagonal passes over the surface, and then push the sanding block back across the area turned approximately 90 degrees from the first pass. This helps to ensure that the entire area is leveled and smoothed evenly.

## Getting It Flat and Smooth

Whether you are repairing collision damage, patching rust, or merely prepping for paint, your objective is to get the panels as straight and flat as possible. This does not mean that you are trying to remove original body lines. You are simply trying to make the surface as smooth and blemish free as possible. We have all seen natural lakes or manmade reflecting pools that mirror the image of the surroundings. Being liquid, water in these lakes or pools is "flat," allowing it to reflect so

No, this is not trick photography. The hood side and cowl (as well as the rest of this '32 Ford) are that reflective. This is what is referred to as "flat." Just as with water in a reflecting pool or a quiet lake, any waves or ripples would distort the reflection of the surroundings. The cowl and hood side are at slightly different angles, thus the dual reflections of the tire.

perfectly. If you toss a rock or coin in the water and disrupt the flat surface, the reflected image is distorted, even if just temporarily. Having body surfaces that are flat will yield the most brilliant paint and allow for the brightest shine. Any imperfections that you can find now will be greatly magnified after applying a new coat of paint.

Even though this grille shell includes multiple compound curves, it is optically "flat" as well. In this case, the shape of the metal causes that little bit of distortion in the reflection.

To get this flat, reflective surface, move the sanding block in as many directions as possible throughout the entire sanding process. Do not simply move the sanding block in a back and forth direction from the front to the back. Move up and down and crossways diagonally, rotating the board or block as necessary for ease of operation. This multidirectional sanding technique will guarantee that all areas are sanded smooth without grooves or perceivable patterns.

### Guide Coats

A guide coat will help you determine if the panel is as flat and smooth as possible. Prior to changing sandpaper grits, spray a mist coat of paint onto the panel you are working on. Inexpensive, spray can enamel in any color that easily stands out from the primer that you are applying it to works best. You certainly do not need to attempt to cover the primer, just a uniform mist is all that is really needed. Allow the guide coat to dry, but if it takes more than five minutes, you have probably applied much more than you need.

The guide coat will be sanded off prior to additional applications of anything, so a spray can of virtually any quick-drying enamel will work. Use a color that contrasts with the color primer you are using and apply a uniform mist to the pieces being guide coated.

Now begin sanding with the next grit of sandpaper applicable to your place in the paint-prep process. If the guide coat is sanding off uniformly, the panel is fairly straight. If the guide coat comes off quickly in some places, those areas are higher than the surrounding area. If the guide coat does not sand off, that area is lower than the surrounding area.

### Masking

Masking tape will not stick to dirty, greasy, or wax-covered surfaces—it might stick initially, but air pressure from a spray gun will cause the masking tape to quickly lose its adhesion

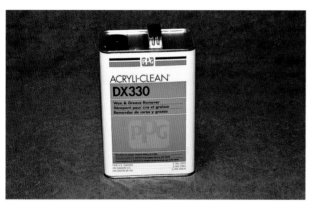

All paint companies have their own brand of wax and grease remover, and there are several independent brands. Regardless of what filler, primer, and paint products you use, be sure to use some wax and grease remover to clean the affected area prior to masking and painting.

Available in different styles, a masking paper tree like this one makes masking a much easier task. As the masking paper is pulled from the tree, masking tape is applied to one edge, making it easier to position and secure to the vehicle. You will still need to manually secure the remaining edges with tape.

Wrinkled and multilayered newspaper is good at trapping dust, too—until pressurized air from your spray gun blows it free and onto your paint coat. Masking paper is available wherever you purchase your primer and paint products and is relatively inexpensive when compared to its ease of use and effectiveness.

Masking paper and automotive-grade masking tape should be used to mask the areas that are in the direct path of the application of primer or paint. To cover larger areas that are not in the direct path of primer or paint, large plastic masking material that is available at retailers of automotive paint products can be used. This material is similar to food wrapping material and is very thin. The thick plastic sheeting typically used as a drop cloth when painting the inside of your house is much too thick and heavy for use on automobiles. Although an initial estimate of the masking needs for your vehicle may appear to be rather limited and easy to accomplish, you should understand that less than meticulous masking will almost always result in obvious spots of overspray, those imperfections that clearly indicate sloppy work or inexperience. Some paint overspray can be cleaned off, but proactive masking is much easier than reactive cleaning.

Painting panels adjacent to windows always presents masking challenges, as there are usually several pieces of molding and trim surrounding the glass. While the most efficient way to avoid errant application of paint is to remove the trim, molding, and glass, that is not always a practical solution. If these areas must be masked, do yourself a favor and purchase a roll or two of ⅛-inch Fine Line masking tape in addition to the automotive-grade masking tape and masking paper that will be required for the job.

Outline the area to be masked with ⅛-inch Fine Line tape, making sure that this tape covers the very edge of whatever is being masked. Now you should attach ¾-inch automotive-grade masking tape to the Fine Line tape. For

Like masking tape, Fine Line tape is available in different widths. Unlike all but the narrowest of masking tapes, Fine Line is much more flexible, allowing it to follow complex shapes more easily. It is also used for laying out flames, scallops, and other graphics.

capability. So, before you begin masking, be sure to use wax and grease remover to clean those areas where tape will be placed.

Prior to spraying any primer-surfacer to areas that have been repaired or modified, mask off any surface that is not to receive primer. Although many a driveway paint job has involved masking with newspaper, it's risky. Newsprint does not hold up well to the solvents in paint and primer materials. Also, to avoid soak-through, you must typically use several layers, which makes taping it down to stay difficult.

narrow areas, such as window trim, one or two additional strips of tape will probably cover the area to be masked. If you are masking a large area, secure masking paper to the first piece of ¾-inch masking tape. Be sure to use masking tape to cover any seams in masking paper.

When masking window glass, only one piece will be necessary as long as the masking paper is wide enough to reach from the top to the bottom of the window. Fold the paper as necessary so that it fits neatly along the sides of the glass. Use strips of masking tape to hold the masking paper in place. If the glass you are masking is wider than the masking paper, use two or three strips of masking paper placed horizontally to cover all of the glass. Lightly secured paper edges will blow open during spray paint operations and allow mists of overspray to infiltrate spaces beneath paper. Therefore, always run lines of tape along the length of paper edges to completely seal off underlying areas. This is especially important when the edge of one piece of masking paper is lapped over another.

Do not forget to mask doorjambs before spraying primer and paint. You would be surprised how much paint can find its way between a closed door and the doorjamb when you do not want it to. If your bodywork calls for priming and painting door edges and doorjambs, these edges should be painted first and allowed to dry. After they have cured sufficiently for masking tape and paper to be applied, mask off the inside portion of the door and the doorjamb. The door can then be closed and the exterior primed and painted.

This same procedure should be used if you are replacing the doors. It is much easier to paint the interior side and perimeter edges of doors while they are off the vehicle. The doors can then be installed on the vehicle, the door edges and doorjambs masked off, and then the exterior panel painted along with the rest of the body.

Two-inch tape and 4- to 6-inch paper can be used to mask doorjambs and edges; however, consideration must be given to the location of the tape edges. If they are set too far out, they may allow a paint line to be visible through the gap between the door and the jamb. This is an important factor when painting a color on the exterior that is in contrast with the shade on the doorjambs. You will have to decide where the dividing line will be and make sure that tape is positioned symmetrically.

Always inspect your masking work after tape has been positioned. Use a fingernail to guarantee tape is securely attached along edges. On the bottom sides of the body, you may have to lie down in order to accurately place masking tape. When you are spraying paint is not the ideal time to realize that masking tape is coming loose or was not located correctly.

## UNDERCOATS

Undercoats should not be confused with the thick, black, sticky tar substance that is applied to the underside of new vehicles to minimize rust through. For the purposes of auto painting, undercoats are the necessary substrates of various primers that precede the application of paint and other topcoats. Undercoats provide basic corrosion protection to bare metals, increase adhesion of whatever substrates or top coats follow, and provide a surface that can be sanded smooth. Since undercoats in general perform a multitude of tasks, specific types of primers must be used for each step of the priming process.

### Cleaning

Paint and primer products simply do not adhere well to dust, dirt, grease, or other contaminants that are often around automotive vehicles. After sanding, whether it is with 36-grit, 3,000-grit, or somewhere in between, you should use an air hose to blow away any residual sanding dust. Be sure to use air pressure to remove dust from between door edges, doorjambs, hood, and deck lid edges.

When all of the loose particles have been blown away, use a liquid wax and grease remover product to thoroughly wipe down and clean the surface prior to spraying any primer. Each paint manufacturer has its own brand of wax and grease remover that constitutes part of an overall paint system. To be sure that the cleaner product is compatible with the paint products, you should use only products from the same manufacturer.

Dampen a clean cloth (heavy-duty paper shop towels work great) with wax and grease remover and use it to wipe off all body surfaces in the area of expected paint undercoat applications. The mild solvents in wax and grease removers loosen and dislodge particles of silicone dressings, oil, wax, polish, and other materials embedded in or otherwise lightly adhered to surfaces. To assist the cleaning ability of wax and grease removers, follow the damp cleaning cloth with a clean, dry cloth in your other hand. The dry one picks up lingering residue and moisture to leave behind a clean, dry surface. Use a new towel on every panel, wipe wet, and *dry* thoroughly. Another method is to dispense wax and grease remover from a spray bottle, and then wipe it off with a clean cloth. (Again, heavy-duty paper shop towels work great.)

### *Epoxy Primer*

Epoxy primer (a.k.a. etching primer) should be used as the first undercoat atop bare metal, whether it is new metal construction or vintage tin that has been stripped of paint and other primers. Two main reasons for using epoxy primer are its superior corrosion protection and its excellent adhesion qualities. Any time that a sheetmetal panel is stripped (chemically or mechanically) to bare metal, it should be cleaned and coated with epoxy primer as soon as practical to avoid the formation of surface rust. Most body fillers can be applied over epoxy primer, so there will be no reason to go back to bare metal during the repair process.

With all primer and paint products, check with your source for the correct components to be used for the parts and pieces being coated and for use under your shop conditions. While the epoxy primer is the same, the two catalysts shown will provide different characteristics.

Most epoxy primers instruct the user to scuff the surface prior to additional coats of primer or paint if the original epoxy primer has been on for more than a certain length of time.

Although this truck cab is in epoxy primer at this point, the reason for the photo is the body dolly. This one appears to be a welded steel frame with a caster at each of four corners. Requiring just three or more casters and some scrap steel or lumber, a body dolly underneath your hot rod project makes it easier to maneuver in the shop.

Be sure to position parts to be primed or painted so that the entire surface is accessible and that lighting is at its best. This folding metal sawhorse appears to work well with this truck fender.

### Filler Primer/Primer-Surfacer

Commonly referred to as filler primer or primer-surfacer, this product should be used as bodywork is being done. Most, if not all, of this will get sanded off as the body is block sanded. An application of primer-surfacer will quickly provide you with visual evidence of how complete your bodywork has been done. What may look great when still in bare metal or body filler will show every low spot, high spot, or other blemish when a coat of primer/surfacer has been applied. However, this is a good thing, as it is much better to gain a better concept of the bodywork shortcomings now, rather than later. Although primer-surfacer is often referred to as high build primer, it should not be used as body filler. If there are indeed low spots that need to be filled, additional body filler should be used.

If you have not done so since the last bodywork or sanding has been done, use an air nozzle to blow off any loose dust and dirt from the surface. Clean the area with wax and grease remover, and then use a tack rag to remove dust particles and lint. Be sure to mix the primer-surfacer and reducer per the mixing instructions on the label, and then apply two or

After scuffing the epoxy primer that had been on the track T for several months, it was time to spray all of the sheet metal (actually fiberglass for the body and aluminum for the hood) with primer-surfacer. Note the use of a paint suit, mask, and a partially open garage door.

Regardless of how long or short the fabrication process has been, you have surely reached a milestone when the body and related components are in one color of primer. It is not ready for paint yet, but it is looking better all the time.

three coats of primer-surfacer to all areas where bodywork has been done. Make sure that you allow the proper flash time between coats. If you have several localized spots of repair on one or adjacent panels, go ahead and apply primer-surfacer to the entire panel rather than just the spots where repair work has been done. This will allow you to better blend the surfaces during the block sanding process.

After the primer-surfacer has had ample time to cure, spray a light, but uniform mist coat of SEM's Guide Coat (or any contrasting color spray can enamel) onto the areas to be sanded. After this guide coat dries, break out your favorite sanding board with some 320-grit sandpaper and block sand the entire area where primer-surfacer has been applied. The guide coat will quickly disappear from high spots, but not from low spots.

After this initial sanding with 320-grit, it should be obvious if any additional filler is required. If filler is required, add it as needed, shaping it with 100- and 240-grit sandpaper. Then apply two or three coats of primer-surfacer to all areas where additional filler was applied. Block sand these areas again with 320-grit sandpaper. If the body is at the

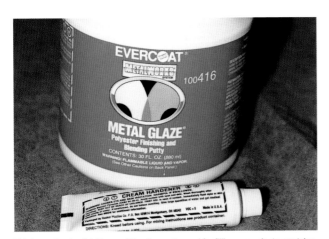

This is a polyester finishing and blending putty used for filling very minute scratches in bodywork. It mixes with hardener, much like body filler, but is much finer.

proper contour after sanding with 320-grit sandpaper, switch to 500-grit sandpaper and block sand the entire area again.

Sanding with finer sandpaper grits will focus on texture smoothness and removing sanding scratches and very shallow imperfections. Using the longest sanding board or block available with progressively finer sandpaper is the key to achieving the straightest and flattest surface upon which to apply paint.

### Sealer

Sealer is typically the last of the undercoats to be applied before color. As such, all bodywork should be completed with all sanding done. The purpose of the sealer is, as the name suggests, to seal in all of the various undercoats and fillers, to keep them from seeping into the top coats. The sealer will also provide a uniform base for uniform color coverage and aid in adhesion of the top coats.

Before applying sealer, use an air hose to blow dust and dirt out of all cracks and crevices. Clean the surface with wax and grease remover, and mask off all areas that should not receive sealer. Blow off the surface again using an air hose, clean again with wax and grease remover, and go over the area with a tack cloth. Mix the sealer by following the directions, set the air pressure as directed, and wear your protective gear. Spray sealer first over the areas where filler has been applied, then feather the sealer toward the masked edges, but do not spray all the way to the edge. Allow the sealer to flash dry per the directions, and then apply second and third coats with the appropriate time between coats.

Sealer should not be sanded, unless runs or other imperfections occur. If that happens, allow the sealer to dry, use fine-grit sandpaper to remove the blemish, and touch up with more sealer. Once the sealer has cured according to label directions, paint can then be applied.

### Seam Sealer

One of the reasons that vehicles rust is that many adjacent panels are simply plug welded together at the factory. A way to avoid this when building a hot rod is to stitch weld all of the sheetmetal panels together. This would greatly minimize the potential, as there would be no open seams to attract and accumulate dirt, debris, and moisture. If you are not actually replacing any panels or complete welding is not feasible, you can apply new seam sealer.

If the components to be sealed are going to be stripped (mechanically or chemically), that should be done first, and then epoxy primer applied. Any existing seam sealer should be removed prior to installing new seam sealer. To receive seam sealer, the metal surfaces should be clean, but it is not necessary for any primer or paint to be removed. Seam sealer is available in a tube that can be dispensed with a caulking gun or in a tube that can be squeezed by hand. In either case, simply apply a bead along the seam between two adjacent panels.

# PROJECT 11
# Block Sanding and Applying Sealer

My current hot rod project is to complete the '27 Ford track T roadster that was the basis of my book, *How to Build a Cheap Hot Rod*. With that hot rod running and drivable, it was time to finish the required bodywork, block sand to achieve a suitable finish, and then apply sealer. After countless hours of bodywork, applying sealer is a major milestone in any automotive project, as it signals the beginning of the painting process.

For a high builder primer surfacer, I used PPG's Omni™ MP282 mixed with MH283 hardener and MR reducer at a 4:1:1 ratio. When the bodywork was completed, I mixed PPG's Omni™ MP172 (black) epoxy primer with MP175 for

use as a sealer. Sealer is used to seal off the body filler and primers used previously and provide a uniform colored base for the paint that will be applied next.

A project the size of the track T is a good first-time car painting project, as it is not so large that it will overwhelm you, yet it provides an introduction to many of the obstacles that you would face on any other auto painting project. Just as applying sealer closes a chapter on the bodywork portion of your hot rod project, it closes this chapter in this book. In similar fashion, "Base Coat to Pinstripe" will be documented in the next chapter. Hang on, it's gonna be a wild ride . . .

**1**

Epoxy primer has been on the track T for several months; therefore, it has to be scuffed up before any additional substrates are added. Scuffing the body also serves as a good opportunity to check for and note any areas that may need additional attention.

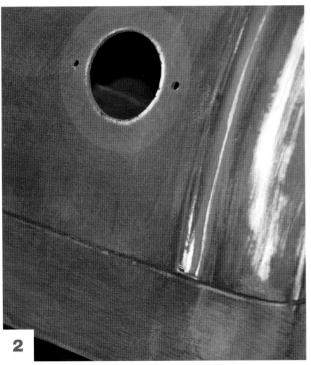

**2**

At about seven o'clock from the passenger side taillight, there was a small (approximately dime size) indentation in the fiberglass. This may have been a defect in the mold, a dent encountered during shipping, or something bumped against the body during construction. Regardless of its source, it needs a bit of filler and some sanding to make it go away. I circled it with a permanent marker for future reference.

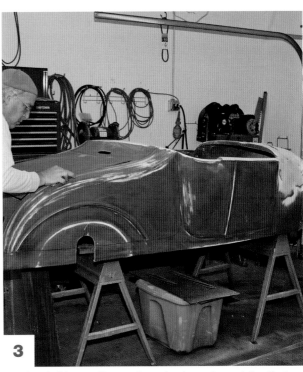

**3**

To do the paint job preparation correctly, you have to spend as much time, if not more, tending to the areas that contain body lines as you do open areas.

**4**

For sanding large, flat areas, a long board sander cannot be beat. Be sure to sand in all of the curved and recessed areas as well, even if it means using a different sanding block.

**5**

These fiberglass body aprons serve to cover the frame rails of the track T, so their underside will not be seen by anybody, unless the vehicle is on a lift rack. Still, they should be given the very same amount of detail as the topside of the vehicle.

**6**

The aluminum hood sides and their louvers need to be sanded as well. To ensure that I did not exert too much pressure and bend the hood sides, I supported them with a couple of scrap pieces of oak flooring while sanding.

**7**

When it seemed like all bodywork and surface imperfections had been addressed and two coats of primer-surfacer applied to the various pieces to be painted, a guide coat was added.

**8**

As the guide coat is sanded, areas where it comes off quickly are higher than surrounding areas. Areas where the guide coat lingers are lower than surrounding areas. Therefore, guide coats should be consistent and uniform to provide the best results, even though this hood top is not a perfect example.

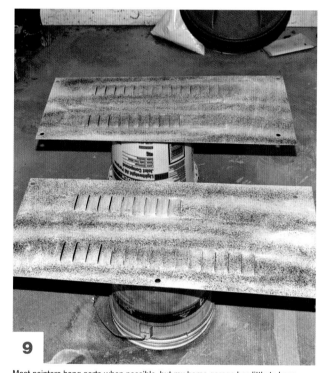

**9**

Most painters hang parts when possible, but my home garage has little to hang parts from (a fact that will be addressed prior to the next painting project). In the interim, a couple of empty drywall mud buckets were used to support the hood sides during the priming and painting. An important thing to remember is to make sure that you do not drag the air hose into or across the parts and that you do not bump into them yourself.

**10**

No fancy downdraft spray booth being used here. Just an ordinary, two-car attached garage that serves to house my wife's car and my hot projects on any other day. The walls are painted white to reflect the small amount of light there is. Note that the electric heater was not running when spraying was being done.

**11**

To protect my photo equipment, I did not take any photos while spraying, but this is what everything looks like after the first coat of sealer. You will notice that the floor is considerably darker after spraying the black sealer. Hindsight being 20/20, I now see that I should have placed some drop cloths over much of the stuff in the garage.

**12**

After a couple of coats of sealer, this is what the hood top and nose look like. For the sealer that I was using, color must be applied within 72 hours, or the surface will need to be scuffed and reshot.

**13**

It was awfully tempting to just stop after applying the black sealer, which is what many rodders have done. How much detail you go into when finishing your hot rod is purely up to you. Going past this point was going to pull me out of my comfort zone, but as you'll see in the next chapter, I now have that first paint job under my belt.

# Chapter 7
# Painting: Color and Other Top Coats

Now that you have a hot rod chassis that has the perfect stance, and the fenders, doors, hood, and deck lid have consistent gaps, and the sheetmetal and/or fiberglass body components are arrow straight and silky smooth, it is time to lay on some color. If you have done everything up to this point (and your quality has proven acceptable), there are few reasons why you should not paint your hot rod yourself. Not that it should be taken lightly, or just anybody off the street can do it well, but the actual practice of mixing some paint pigment, reducer, and hardener together, pouring the concoction into a spray gun cup, and pulling the trigger is fairly simple.

All paint preparation jobs are related to a single goal, a perfect paint job. However, each has its own function, which will not improve on another's lack of perfection. So, take your time during each preparatory phase and do not go on to the next step until satisfied that the work you just did has been accomplished accurately and completely. If you are feeling hesitant about your bodywork or paint prep, go back and redo anything that you did not get quite right.

Prior to adding any color to your hot rod project, considerable thought should be given to two things: color (or colors) and paint scheme. Besides the actual application, these two things will have the most influence on the overall appeal of your hot rod to others. Of course, your primary goal when building a hot rod should be to please yourself, but if it is well received by others, that makes the extra effort worthwhile.

## COLOR

Lots of colors look great on hot rods; red and black are just two of them. There is no denying that red and black have been popular with hot rodders for a long time, and most likely always will be. Fire trucks, Ferraris, and countless other vehicles are red for good reason. Emotionally, red captures our attention, allows us to lose our inhibitions, and generally makes us feel good. Realistically, red being a bright color makes it relatively easy to keep clean, it photographs well, and it is a bit more forgiving than certain other colors if the bodywork is a nick below perfect. Black is popular for some of the same reasons as red, but for some that are somewhat the opposite. A black paint job is difficult to keep clean, can be a real bear to photograph because it absorbs so much light, and requires nothing less than perfect bodywork, as any flaws will seem to be magnified. Yet, when a flawless bodied hot rod is coated with the slickest of black paint jobs, nothing looks better, nor shows off the artist's work as well.

While this black roadster would be as much fun to drive and would be built to the same degree of quality in just the black paint, the hot licks wake it up substantially. The magenta flames surely gain this deuce some notoriety, but that may or may not be what you want in a hot rod.

Some rodders would be content with a simple, yet well-executed black paint job, with just a bit of pinstriping. If you don't think this rod is flat, look at the reflection beneath the rear window.

Still, there are other colors—some are timeless, and some are quickly outdated. Also popular are neutral colors that are somewhat similar to the factory colors from when our hot rods first rolled off the assembly line. Light and dark tans, most shades of gray, bluish gray, and light olive green solid colors seem to be popular and will most likely look tasteful throughout the lifetime of the paint job itself. Even if the color selected is not a period perfect color match to an original, the muted color seems to transcend time.

While it may or may not be intended to be an exact match to an original Ford Model A color, it has to be close—sort of like taking a brand new Model A, stripping off the fenders, running boards, and hood sides to build a hot rod. Isn't that how it all started?

The metallics that are often found on higher-end OEM vehicles are well received, too. When these colors are used on hot rods, they are commonly used with a two-tone paint scheme and are typically paired with a neutral color. While the basic colors cover a wider portion of the spectrum—green, blue, orange, and some reds—these colors are still being muted substantially. In other words, we are not talking emerald green, sky blue, or fleet color orange. Like the OEM vehicles that originally utilized these colors, most of the hot

rods using them are being built with all the stops out. Like wearing a black tuxedo and a pair of tire tread sandals, some of these high zoot paint schemes simply are not gonna look right with painted steel wheels and hubcaps.

This pewter-colored sedan delivery is a common example of the high quality hot rods being built. Where loud and rough around the edges was the norm years ago, polished, refined, and sophisticated is now the style of the average hot rod. The smooth bodywork, slick paint, and subtle chrome appointments are right on par with any luxury car.

Adding a bit of color, albeit subdued, along with some muted tones in betwixt the two colors looks great on this coupe. You most likely would not choose these colors just by looking at a color chart in a paint store, but looking around at late model vehicles, you will find these types of colors everywhere.

As hot rods are often built to say "look at me," it is no surprise that some rodders choose colors that are bold and bright, rather than the subtle examples mentioned thus far. A good suggestion when trying to choose a color is to simply look at other vehicles when you are in traffic. What OEM colors tend to look good whether the sun is shining or it is a cloudy day? Which ones lose their appeal after they have been on the road a while, and which ones turn your stomach? Having the actual paint code from the vehicle will help

While this two-door sedan features a paint job that fades from white to very dark blue, the steps in between are very broad, with little area used to blend. Rather than use specific colors to paint the different bands of color, a more effective approach may have been to simply use the white and the very dark blue in graduated amounts (or coats). In reality, that may be more difficult than it sounds.

considerably, but if you can at least determine what make and model the vehicle is, you can most likely find out which color it is when you go to the paint store. If you find the color that you like, but still are not positive that it is the color you want on your hot rod, purchase a pint and spray some sample panels to see if it is what you really want.

## PAINT SCHEME

Besides choosing what color to paint your hot rod, you will also need to decide on a paint scheme. If you decide on a one-color paint scheme now, you can always add flames, scallops, or other artwork to it later if you choose. However, if you know that you want something other than just one color, it may be easier to go ahead and add it now rather than take the vehicle off the road for some length of time.

For any sort of paint scheme that involves multiple colors, you should do some preliminary design work ahead of time, rather than just whipping out the masking tape and proceeding haphazardly. Whether you sketch out your vehicle on a drawing pad, work out your creation on a computer, or peruse stacks of magazines and/or photographs, it is good to have some sort of visual guideline as to what you want to reproduce. Shapes of flame outlines and scallops are more difficult to layout than one might think, while attempting to do it without any visual reference is simply making the task harder than it needs to be.

While the processes are the same, realize that anything other than a monotone paint job will require additional work. Except for detailed custom airbrush work, each color of paint that is sprayed will require that the entire vehicle be masked off, except for the area being sprayed at the time. If you have masked the vehicle properly one time, you can do it again; however, this may be more than what you care to try if this is your first time painting a vehicle.

When applying multicolors, it is even more critical to observe the proper flash and drying times for the paint system you are using prior to masking. Applying masking tape or masking paper to freshly painted surfaces that have not yet dried adequately can thoroughly make a mess of your freshly applied paint. Product information sheets for the specific type of paint you are using will provide a specified time to allow the paint to dry before taping. Likewise, clear coats (if applicable) must be applied within a specified time, or the base coat will need to be scuffed and additional base coats added.

### One Color or Monochromatic

If a hot rod is painted just one color, but uses chrome trim, bumpers, and mirrors, is it monochromatic? Back some

If you are going to be utilizing all of the stock trim (or even custom trim), you must realize that the trim itself serves as a graphic. Therefore, you would not want to use multiple colors on this hot rod that has an abundance of stainless trim. It would be easy to two-tone portions of the car with the break being under the trim, but there would be large areas where the different colors would be right next to each other.

Painted steel wheels always seem to look better when they are a different color than the body. This is especially good to remember if you have the wheels powder coated, as you most likely would not be able to match the colors anyway. Other than an approximate 2- or 3-inch top chop with raked back windshield posts and trim removal, this truck body is pretty stock . . . and there is nothing wrong with that.

number of years ago, hot rodders were painting almost everything (wheels, exterior trim, etc.) on their hot rods the same color as the body. Magazine scribes quickly began referring to this craze as monochromatic, which technically is correct. Like most any style, monochrome looks better on some vehicles than on others. While trim can be repaired and rechromed, being able to fill small imperfections with a dab of body filler and paint is an easier and less expensive alternative. With that said, any additional references to monochrome or monochromatic in this book simply mean that something is painted one color, regardless of the finish on exterior trim.

### Multicolor

For a simple one-color paint job, you can clean the vehicle with wax and grease remover, mask off everything that isn't going to be painted, remove any dust with a tack cloth, and

The two-tone paint job on this sedan delivery breaks the paint at a logical location, just below the lower window line. Breaking the paint higher would interfere with the available "canvas" that would be glass on a regular sedan. Breaking the paint lower would visually lift the tops of the fenders.

then apply paint. If multiple colors are going to be involved, it starts getting a little more complicated . . .

On most production vehicles that have a two-tone paint scheme, the color is usually broken at a body line, which makes masking easier. Additionally, there is usually some sort of trim that covers the edge of the different colors. On your hot rod, where should the primary color stop and the secondary color start? Is there a piece of trim that covers this seam and will that trim piece still be used after the paint job? Does the trim cover the entire paint seam or is some of it left out in the open? Which color are the doorjambs? How well you define color transition will have a great impact on your overall paint job. It is better to have a high-quality monochrome paint job than a mediocre two-tone finish.

When multiple colors (two-tone, flames, scallops, etc.) cross a door or deck lid opening, the painter must determine a suitable transition. Most painters will continue the graphic onto (and all the way across) the doorjamb, as done on this coupe. Remember that the doorjamb is basically a continuation of the exterior of the door, until it reaches the upholstery on the door panel or kick panel.

Think about how you would mask the area that is to be a different color and realize that it should be the same on both sides of the vehicle. This will help you realize that you need to take advantage of body lines and natural breaks if you are designing a two-tone paint scheme. If you are going to be hiding a paint seam under a piece of trim, be sure to split the width of the trim evenly with each color. If the trim is an inch wide, this leaves a half-inch for each color. You should be able to align the trim accurately enough to cover this; however, if the trim is narrower, you will have less room for error in masking or trim installation.

Although I admittedly do not recall for sure, I would bet that all of the hood top of this deuce roadster is black, while the hood sides (if there are any) are painted the silver/pewter color in their entirety. Taking advantage of natural body line breaks makes masking easier and usually leads to a more pleasing paint scheme as well.

It does not really matter which color of a two-tone paint scheme you paint first; however, you should plan ahead to make your masking work easier. When all of the bodywork is done, sealer is applied, and all areas that are not to receive any paint masked off, no additional masking would be necessary for the first color to be applied. After this first color is applied, you will need to mask it off too before applying the second color. For this reason, it makes sense to spray the color to the area that is going to be easiest to mask first, whether it is the lighter or darker color.

## TYPE OF PAINT

A wise hot rodder friend of mine has mentioned to me on more than one occasion that rodders really need to find their

With all of the magazine articles that have been done on faux patina paint jobs, and the fact that this pickup is from the Midwest, it is hard to say if the surface rust is real or not. Who knows if the tailgate is out of a salvage yard or a brand new piece.

level of comfort when it comes to hot rods. Many rodders try to outdo each other when they are building or updating their vehicles. Whatever one does, the others do. That is OK if you can really afford it, but many cannot, and they end up getting into hard times. Although that applies to the entire hot rod, it probably becomes more evident on the paint than any other aspect of the vehicle.

Although paint application basics remain at least somewhat constant, the types of paint available have greatly changed over the years. The cheap and easy-to-use acrylic lacquer paint from a few decades ago is now a thing of the past. Likewise, plain old "one coat" acrylic enamel is gone as well. Even though those products were easy to use, their replacements are far superior in finish quality, durability, and user friendliness.

A little too shiny for primer, a little too flat for glossy. . . . While this could be an application of sealer prior to applying finished paint, I would imagine that it is finished paint that has had flattener added to provide the semigloss appearance. Paint is really just a method of protecting the surfaces of a hot rod, so it does not have to be glossy if you don't want it to be.

### Single Stage

Urethane enamel paint products (a.k.a. single stage) consist of a color pigment (paint) to which a prescribed amount of hardener is mixed just prior to application. Two or three coats of urethane enamel on a properly prepared surface will yield a high-gloss finish without buffing and will be very durable. A downside is that there are fewer layers of material between your vehicle's body and the sometimes harsh elements of the world. In other words, if the paint layer does get scratched, it could easily expose the primer or even the bare metal, making it susceptible to the formation of rust.

Some painters believe that a single stage paint system that is wet sanded and buffed yields better color. This may very well be true, as many people feel that the clear of a base coat/clear coat system takes away some of the appearance of depth that was common in the old lacquer paint jobs. A downside of a single stage paint system is that the layer of paint on the vehicle is going to get thinner each and every time that you rub a car washing sponge, chamois, or wax applicator across it.

### Base Coat/Clear Coat

If you are painting anything other than a one-color paint job, you will make your job easier by using a base coat/clear coat paint system. Using a base coat/clear coat system will allow you to apply two coats of clear over the first color in a multicolor paint scheme. If you inadvertently miss any of the first color when you are masking, slight overspray from the second color can be sanded out of the clear that is protecting the first color. After the second color (and third, if applicable) is applied, the entire vehicle should be coated with two or three additional coats of clear.

Whether one color or several, a base coat/clear coat system involves more spraying time as you must apply the base coat in enough coats to obtain coverage and then apply the desired number of clear coats. You must allow the proper flash time between coats, and you must also wait the proper amount of time for the base coat to dry before applying clear. Product information sheets for your specific paint products provide all of the flash time, dry time, and application pressure information that you need. You just need to make sure that you ask for them when you purchase your paint products.

After you apply clear and it has had adequate time to dry, it can be wet sanded to remove any surface imperfections. Removing these imperfections will make the surface more optically flat, which is what provides the basis for the ultimate shine.

### Tristage

Originally reserved for custom paint finishes or high-end vehicles, tristage paint systems are becoming common on OEM vehicles. By combining a base coat, a color coat, and a clear coat, some exotic colors can be created, including pearls and candies.

Base coats are usually a metallic color such as gold or silver, but can also be black or white. The color coat can be virtually any color, but it will vary in appearance, depending on the base color to which it is applied. For example, a red color coat sprayed over a silver base will yield a different tint than if applied to a gold base coat. The clear coat finish will prevent wet sanding or polishing from distorting the blended color achieved between the base coat and color coat.

As you are applying multiple coats of three different products, the amount of time required when using a tristage paint system increases considerably. They are not recommended for the beginner painter. Yet if you are experienced with multistage paint systems, the possibilities in colors and custom effects are virtually endless.

### Waterborne

Where conventional paint products use solvents to suspend the pigment (make it liquid), waterborne paint uses water—similar to oil-based versus latex house paint. The bad thing about paint solvents is that they contain volatile organic compounds (VOCs). VOCs are chemical substances that rise into the atmosphere from paint overspray and solvent evaporation to unite with nitrous oxides to produce ozone, which is a major component of smog.

To minimize VOCs and keep us healthy, paint manufacturers are researching and developing waterborne paint products. In the state of California, some municipalities throughout the United States, and portions of Europe, this new paint technology is currently mandated by law. It most likely will become widely used throughout the United States before the end of the first quarter of this century. Paint spray guns that use aluminum or stainless steel internal components to avoid rust are readily available, even though the waterborne paint products are not. In all reality, waterborne paint products will not be available in your local area until just before they are required by law.

## After Applying Paint

Now that the paint has been applied, you are much closer to being finished than when you first started, but there is still more work to be done. First, you will need to thoroughly clean all of your paint spraying equipment so that it is in good operating condition for the next time you want to use it. Secondly, you will need to carefully remove masking material from the vehicle after the appropriate amount of drying time. To achieve the finest quality finish, you will also need to wet sand away any slight imperfections from the paint and then buff it to a high gloss.

### Removing Masking Material

As anxious as you will be to see what your vehicle looks like with its new paint, you must show some restraint as you remove the masking material. If you simply tear off the masking paper and tape with reckless abandon, you will no doubt do at least some damage to the new paint in the process. The reason you need to be careful is that some amount of paint (color or clear) will be overlapping the edge of the masking tape. If you pull the masking tape straight up from the painted surface, the paint will tend to flake along the edge. To prevent this from happening, pull the tape away from the freshly painted surface and back upon itself so that the tape is leaving the surface at a sharp angle.

If several coats of color and clear have been applied to the body surface, it may be necessary to cut through the layers of paint manually at the edge of the tape. If you need to do this, use a sharp razor blade, take your time, and be careful. You do not want to damage the paint or slice open an appendage.

### Wet Sanding

After the paint has had sufficient time to cure, it's time for wet sanding (or "color sanding"). This step will remove any texture in the paint and yield a mirror finish. Before wet sanding, confirm with your paint supplier that the paint you bought and sprayed can be wet sanded. Urethane enamel can be wet sanded and buffed, just like clear, yet many of the older enamel products should not be wet sanded. As a basic rule of thumb, most paint products that are catalyzed—hardener is added when the paint is being mixed with reducer just prior to application—can be wet sanded and buffed. As some products can be buffed after just 24 hours and others require 90 days, you should verify the necessary requirements with your paint supplier.

Base coat/clear coat or tristage paint systems are the typical candidates for wet sanding, as the sanding can be done on the clear rather than actual color coats. No sanding should be done to color coats in these systems, unless it is to remove runs, drips, or other errors. Note that sanding out these defects may necessitate repainting an entire panel. Especially with candy finishes, sanding directly on the color surface will distort the tint to cause a visible blemish. Your wet sanding efforts should be concentrated on clear coats in order to not disturb the underlying color coats. Wet sanding clear coats will bring out a much deeper shine and gloss when followed by controlled buffing and polishing.

Let the water sit in a bucket overnight, so that any minerals in the water that could cause scratches will settle to the bottom. Use very fine 800- to 3,000-grit sandpaper with plenty of water to smooth or remove minor blemishes. Add a very small amount of mild car washing soap to the water bucket to provide lubrication to the sandpaper and let the sandpaper soak in the water for about 15 minutes before beginning the wet sanding. As with all other sanding tasks, you should use some sort of sanding block. Instead of using a large sanding block, fold sandpaper around a wooden paint stir stick to more accurately remove small nibs of dust or dirt. Use a very slight amount of pressure and be sure to dip sandpaper in a bucket of water frequently to keep the paint surface wet and reduce the amount of material buildup on the sandpaper.

### Clear

Prior to adding any coats of clear, make sure that the color coats are as perfect as you can get them, as clear is not going to hide any imperfections. Do whatever masking is required. Unlike collision repair to daily drivers, hot rods are often painted in several pieces and then assembled after all paint steps are finished, so masking may not be required. Position the parts to be cleared in your spray booth or garage so that as much light as possible is available. Without the colored pigment of paint and in less than perfect lighting conditions, clear is more difficult to spray evenly. Extra thickness of clear is OK, but if you miss a spot, you will quickly damage your paint when you begin wet sanding or buffing the clear.

As a minimum, you should apply three or four coats of clear if you plan to wet sand or buff it afterward. For the greatest gloss and shine, you can apply two or three coats of

clear, wet sand that with 800 to 3,000-grit sandpaper, then apply two or three more coats of clear, followed by more wet sanding.

### Buffing

Just as with wet sanding, you should first check with your paint supplier to verify that the paint system you are using is compatible with buffing and what the supplier's specific buffing instructions may be. In the case of single stage urethane, buffing new paint with gritty compound will actually dull the surface and ruin the finish. On the other hand, base coat/clear coat or tristage systems greatly benefit from buffing, resulting in a more brilliant finish with a much deeper shine even than similar products that are not buffed.

Some buffing compounds are designed for use by hand, while others are designed to be used with a buffing machine. There are also different types of buffing pads, mainly foam or cloth. Some compounds should be used with a foam pad and a higher speed, while others call for a wool pad and a slower speed.

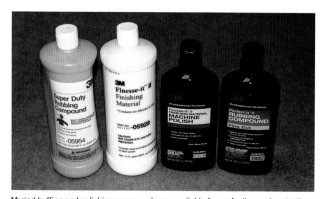

Myriad buffing and polishing compounds are available for perfecting and protecting the paint on your hot rod. Each has its own specific uses and recommended instructions, so it is best to check with your paint supplier to determine the best compounds for your particular application.

Rubbing compounds are made of relatively course polishing material and are designed to quickly remove minor blemishes and flatten the paint finish. Flatten in this instance means to make the painted finish as perfect as possible so that it is more reflective. Since this rubbing compound is coarse, it will leave light scratches or swirls on the painted surface. A much finer buffing compound should then be used to eliminate these fine swirls.

As refinish products have changed over the years, so have some of the ideas on buffing. With the new urethane paint products, do your first polishing with 2,000-grit compound, using a foam pad. This usually minimizes swirls and provides a satisfactory finish the first time around. If swirls are still present, go back to an 1,800-grit compound to remove the swirls, then use the finer 2,000-grit again. Older technology and what we all learned in junior high wood shop regarding sandpaper would have said to use the coarse rubbing compound, then work up to the finer stuff, instead of this seemingly backward procedure.

To use a buffer, first spread out a few strips of compound, each about 4 to 6 inches apart to cover an area no bigger than 2 square feet. Operate the buffing pad on top of a compound strip and work it over that strip's area, gradually moving down to pick up successive strips. The idea is to buff a 2-square-foot area while not allowing the pad to become dry of compound. Continue buffing on that section until the compound is gone and the paint is shiny, and then move onto another area. You must keep the buffer moving to keep from burning into or even through the paint. You must also be especially careful near ridges and corners, which concentrate the buffing force on a very small surface area. If you are using buffing compound that can be applied by hand, use a back and forth motion. This will help prevent swirls, and when you are buffing by hand, you should do whatever you can to minimize your work.

You must also remove dried compound from the buffing pad by using a pad spur. Carefully but securely push a spur into the pad's nap while the pad is spinning. This breaks the dried compound loose and forces it out of the pad. Just be sure to do your pad cleaning away from your car and anything else that you do not want covered with compound or pad lint. As when using most any power tool, you'll have eye protection on when buffing and cleaning the pad.

### Graphics, Artwork, and Trim

Most graphics and artwork are applied in a completely separate operation from the actual painting of the vehicle, giving the paint time to cure fully before placing anything on top of it. This prevents the later artistic touches from trapping solvents that need to escape from the paint.

Any time that graphics or any other artwork is going to be painted on your vehicle, a base coat/clear coat (or tristage) system for the entire car will make for easier repair should overspray or other miscues need to be sanded off. Most blemishes can be removed from the clear without affecting the color beneath it. However, if there is no clear to remove overspray from, you may easily sand through the color right into primer.

Before adding any artwork, the entire area should be cleaned with wax and grease remover. Surface preparation will depend upon the media being applied, so read and understand the instructions provided with them.

### Flames

In one form or another, flames have been painted on hot rods pretty much as long as hot rods have been around. In no particular order, flames may be outline, ghost, single color, multicolor, realistic, traditional, tribal, and three-dimensional.

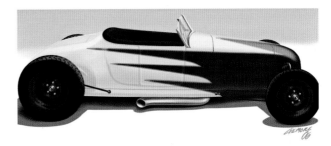

Having a sketch or drawing to refer to when laying out graphics is always a good idea. This particular drawing has the scallops extending straight back onto the turtle deck with an almost vertical line where the colors change. Additionally, the center scallop comes to a distinct point on the nose, just above the headlight. *Artwork courtesy of Stilmore Designs*

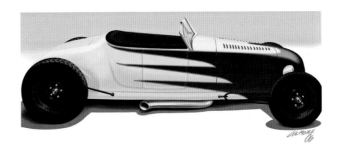

This drawing stops the scallops before getting back to the turtle deck and also wraps them around the curve of the body a little more noticeably. The line between colors is now at more of an angle and somewhat centered between the back of the front tire and the front of the windshield frame. The scallop on the hood also has a rounded belly now. *Artwork courtesy of Stilmore Designs*

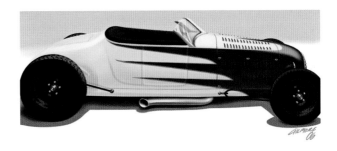

These scallops are virtually the same as the previous. A big difference, and one that emphasizes the importance of doing this on paper first, is that this is where we decided to paint the headlight buckets the dark brown, instead of the light tan. *Artwork courtesy of Stilmore Designs*

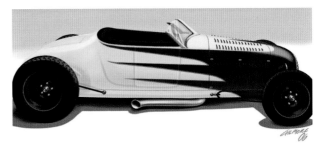

Another subtle difference, and one that will not be noticed by many, is the addition of an orange pinstripe on the painted steel wheels. *Artwork courtesy of Stilmore Designs*

Outline flames are those that have no color of their own, except for the pinstripe that outlines them. The popularity of these seems to come and go, but they are often seen on bodies that are still in primer. This gives the vehicle owner or builder some time to experiment with the shape of the flames with a minimum of fuss.

Ghost flames and single-color flames are essentially the same thing, except for the color or lack thereof. Single color flames have the same color from front to back with no fades, just a solid color that is usually, but not always, in vivid contrast to the main body color. Ghost flames are similar, except that the flame color is just slightly different from the main body color, making the ghost flames difficult to see in some lighting conditions. While ghost flames can be interesting (especially while the vehicle is moving in and out of various lighting conditions), their application requires the same amount of work as those that are fully engulfed.

Multicolor flames are those that involve various colors or shades, such as the traditional red fading to orange, fading to yellow, on a black-bodied hot rod. The area that is not to be flamed must be masked off, but the various colors or shades of paint are usually applied randomly (albeit with an experienced eye) to avoid the appearance of sharp edges.

Traditional and realistic flames are somewhat the opposite of each other. Realistic flames are usually applied with an airbrush and no outline defining masking. What makes them look more realistic is that the outer portion of the flame is brighter than the middle, with the color radiating outward. Traditional flames and their various colors, on the other hand, typically radiate front to back. Even though realistic flames seem more authentic, traditional flames are well . . . traditional.

Flames are not required to be perfectly symmetrical, but they should be balanced from side to side. Flame patterns can be long and narrow, big and bold, cover the bulk of the vehicle, or just a portion. You should lay out the flames so that they do not fight with the natural body lines of the vehicle.

When flames will extend across more than one body panel, all body panels must be assembled prior to the flames being laid out. Even if it is not the final assembly, everything must be in the correct position, even if it means using temporary shims. Doors, deck lids, and hood panels must all be aligned properly so that body lines match up. If this is not done before the flames are laid out, the flames will be misaligned when panel-gap adjustments are made.

Graphics that span multiple panels make preassembly a must prior to laying out the artwork. Having a paint line misaligned by even an eighth-inch between panels will quickly negate any aesthetic benefits of the graphics.

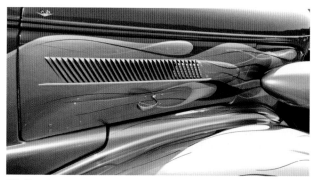

This Chevrolet sedan has several obstacles that would trip up an inexperienced painter. Whoever painted this one seems to have avoided them quite well. Flames were laid out to avoid the louvers in the hood, which saved considerable time and effort. Additionally, breaks in paint color were kept relatively straight where they cross a gap between panels and were also kept around 45 degrees of being perpendicular of the gap.

While the bulk of these flames use the same primary color throughout (albeit with some hints of blue at the ends), they do fade (quite nicely, I might add) from somewhat deep and dark at the front, to a brighter shade toward the back of the car. However, I cannot understand the smaller gold and orange flames that appear just behind the headlight.

To help ensure that the flames are balanced, you should place a strip of ⅛-inch Fine Line tape down the centerline of the area to be flamed from front to back. If the flames will be extending in multiple layers across a large area, such as a hood or roof, this large panel should be sectioned off with tape running from side to side. The length of the flame tips can be balanced in this manner. The tips or bellies of the flames do not have to extend precisely to the guidelines; they're for more general reference on flame length and proportions.

Before beginning the layout procedure, you must decide if you want to merely balance the flames or if you want them to be symmetrical. For symmetrical flames, the flames should be laid out using ⅛-inch Fine Line tape, and the layout completed on just one side of the vehicle. Then lay out a large piece of pattern paper on the side of the vehicle where the flames have been laid out. Make sure that there are no bubbles or pockets in the paper as it is taped down. Mark several reference points on the pattern paper that are distinct locations that can be determined on both sides of the vehicle. Then use a pounce wheel to trace over the flame outline, poking small holes in the pattern paper in the process. A pounce wheel is like a small pizza cutter, except that the cutting wheel has teeth (or spurs), rather than a smooth cutting edge.

When the entire flame pattern has been traced over, carefully remove the pattern paper, flip it over, and position it on the opposite side of the vehicle, using the reference marks mentioned previously. Use masking tape to secure the pattern paper to the body and then spread drafting pounce, carpenter's chalk line chalk, or talcum powder over the paper. Brush the powder into the holes made in the paper by the pounce wheel. Now remove the pattern paper and use ⅛-inch Fine Line tape to lay out the flames by essentially connecting the dots. The holes in the paper will be very close together, but you should still ensure that the tape has smooth curves, even if it means missing a dot or two.

Another method for laying out flames is to lay them out freehand. Use ⅛-inch Fine Line tape and start with the flame that intersects the centerline of the vehicle. Begin at the tip of the flame, working forward toward the belly, and then across the centerline. Then lay out a near mirror image. At this point, you should have the inner boundaries of the first flame lick on each side of the vehicle. The overall style of the flame pattern begins to take shape as the second edge of the first flame is applied. As you progress and the flame layout is cascading across the hood and down the sides, step back on occasion to verify that you like the way the design is proceeding. When you are satisfied that the first side is an acceptable pattern, it can be duplicated on to the other side by using the pattern method mentioned previously or freehand.

Randy Lenger demonstrates the proper technique for laying out some hot licks on the fender of a '40 Ford pickup. This was a quick session for photos; otherwise, he would have waited until the hood was on and would have started in the middle of it and worked his way out. Beginning at the belly, the outline of the first lick is laid out.

Randy then works forward from the tip and then around the next belly. Notice how the Fine Line tape is held in one hand several inches from where the other hand presses the tape down onto the surface.

Randy continues the process of working from the belly to the tips. Starting with a new piece of tape at the tip, around the belly of the flames, and then tearing or cutting the tape at the tip allows the flexible Fine Line tape to be in one piece in the critical belly area. If you were to start a new piece of tape in this curved area, it would most likely pull loose.

Notice also how the bellies of the flames move back on the vehicle as they move outward. This is not an absolute requirement but will depend largely on the overall shape of the vehicle, motorcycle, or refrigerator being flamed.

If these flames were actually going to be painted, masking paper would be used to mask the bulk of the area between and behind the flames, which would remain the base color of the vehicle.

When the layout has been completed, checked, adjusted, and deemed satisfactory, the masking can begin. Overlap the original outline with ¼-inch masking tape, with the excess being applied to the area to be masked, and the original ⅛-inch Fine Line serving as the actual edge of the flames. Use ¾-inch or wider masking tape to mask the remaining area, using straight strips of tape that are trimmed as necessary at the flame outline. Overlapping the masking tape in a continuous fashion will make the tape easier to remove as a sheet later. After masking off the intricate areas within the flame pattern, use masking paper and tape to mask off the rest of the vehicle. Be sure to extend the masking paper down below the bottom of the vehicle to avoid getting any overspray underneath the vehicle.

Use 800-grit sandpaper to scuff the clear that is in the area where the flames will be applied. Blow off all dust and dirt with an air nozzle, clean the surface again with wax and grease remover, and then use a tack cloth to pick up and remove any remaining lint or dust prior to spraying. Now add the color according to the instructions for the particular paint system you are using. Use the proper mixing ratio, proper air pressure, and proper flash time between coats. What now appears as broad strokes of color will be more defined and intricate when the masking material is removed.

After the proper drying time for your paint system, remove the masking paper and tape. Take your time to avoid pulling any paint with the tape. Mask off any areas that are not painted and then apply two or three coats of clear to the entire vehicle.

## Scallops

Generally, though not always, scallops are one solid color. Coupled with the fact that scallops are more of a geometric shape than flames, and therefore easier to lay out, scallops are probably easier for the first-time painter to apply.

OK, scallops are definitely easier to lay out than flames, but that does not mean that they are easy. While piddling around in the garage, I did a bit of practicing with the Fine Line tape and my track T.

One thing that I quickly found out during this exercise is that you might want to use a tape measure to determine equal spacing for the tips of the scallops. If the rest of the car is already painted, you can use a grease pencil to put a dot near where each tip should end, or a small piece of masking tape, whether the car is already painted or not.

The big question for scallops is how many points should there be and should the bellies of the scallops be pointed, rounded, or square with rounded corners? Another layout question is should the tops or the bottoms of the scallop be parallel and to what should they be parallel? These questions are all dependent upon the particular vehicle to which they are being applied. A layout that might look good on an '82 Chevrolet Malibu would not give the same result on a '49 Buick.

While the tips of the scallops on this Model A follow a fairly uniform imaginary line, the bellies are somewhat sporadic in alignment and vertical spacing. Even though the scallops are aligned from the door to the rear quarter panel, the door of a Model A does not fit flush with the body. By spanning these two panels with scallops, the scallops will most likely always appear to be misaligned.

Just as with flames, when applying scallops across multiple body panels, the panels must all be assembled in final position beforehand. Even if it is not the final assembly, everything must be in the correct position, even if it means using temporary shims. Doors, deck lids, and hood panels must all be aligned properly so that body lines match up. If this is not done prior to the scallops being laid out, the scallops will be offset when adjustments are made.

To duplicate the scallop layout from one side to the other, find the centerline of the vehicle. Then run a strip of ⅛-inch Fine Line tape down this centerline from the front to the back of the vehicle. Additional guidelines may be applied in similar fashion to provide reference points from which to measure. As the scallops are cascading across the hood or down the side of the vehicle, they are a repeat pattern, rather than the free-flowing design of flames. When satisfied that the first side is an acceptable pattern, it can be duplicated on to the other side by using the pattern method mentioned previously, or freehand. Mask the scallops, scuff the surface, and paint the scallops in the same way as flames as discussed previously.

Although there is no set rule on the number of scallops, the layout on this roadster is one that has been popular through the years. Just as with any two-tone paint scheme, color choices are important as well. A slightly different shade of red or white could potentially make a drastic difference in the appeal of this highboy.

### Airbrush Work

Custom paint work done with an airbrush is limited only by your imagination as an airbrush artist and your wallet. Before any airbrush work is done, the paint below should be protected by a minimum of two coats of clear. This way, if the airbrush artist were to make a mistake, the work could be repaired without doing harm to the paint below. Likewise, after the airbrush work is added, it should be protected by multiple coats of clear as well.

This coupe displays a variety of paint effects. Obvious is the basic two-tone paint scheme made up of blue over green. The colors break at the belt line and are separated by a tribal flame graphic. The skull is typical airbrush, but larger than life. As the flames erupt out of the skull, they are realistic in color and shape, but then they take on a tribal flame outline.

### Pinstriping

Pinstriping should be used as the finishing touch to any multicolor paint scheme where chrome or stainless trim is not used to separate the two adjoining colors. Pinstriping can also be used as a standalone accent to a single color paint scheme. Since this application is as an accent, a bright color is common, such as an orange pinstripe on a maroon vehicle or purple on a black vehicle. You most likely would not choose to apply a two-tone paint job to a vehicle using equal amounts of these two colors; however, in the thin width of a pinstripe, it all works well.

For painting signs, lettering commercial vehicles and hot rods, and pinstriping, 1 Shot Sign Painters' enamel has been the industry standard for decades. Available in a wide variety of colors, multiple colors can be mixed to create most any color desired.

Graphics such as flames, scallops, or even lettering, simply look, more complete if outlined by tasteful pinstriping. Instead of merely having two colors abutting each other, a pinstripe covers this otherwise unattractive seam of color. It can also serve to conceal slight irregularities in the edge of the paint that may have resulted from the removal of masking material. Whether necessary to hide miscues or not, no flame or scallop painting is considered to be complete if not pinstriped.

Not that there is anything wrong with it and there certainly are no rules, but see how the white outline around the numbers and team name make those items stand out a bit more than the driver's name that is just one color with no outline. Pinstriping around flames and scallops has the same effect.

1 Shot Sign Painters' enamel is the most common paint for pinstriping or lettering and is available from larger art supply stores, some auto body paint suppliers, or through mail order from sources such as the Eastwood Company. It is relatively inexpensive, and with just a few primary colors and some extra mixing jars, virtually any color can be produced. Before applying the pinstriping paint, the surface must be free of any contaminants, so a wax and grease remover is applied with a clean cloth and then wiped off with a second dry cloth. Painted pinstriping can be applied freehand by using a pinstriping brush (a.k.a. dagger) or by using a mechanical device, such as a Beugler pinstriping tool.

of the stripe, and heads with two wheels provide two stripes. The tool can be used freehand, but it will also accept a guide arm that can be run alongside a magnetic guide that can be aligned as desired on the vehicle's body.

### Installing Trim

Prior to installing or reinstalling any pieces of trim that were removed, check to make sure that they are not damaged and are clean and shiny. If the trim is metal, it will be easier to clean and polish if these tasks are done before the trim is reinstalled.

When utilizing stock trim, the key to making it look good is to make sure that it is aligned properly from one panel to the next. This may mean drilling new holes for fasteners or modifying the fasteners to fit the holes properly.

Pinstriping takes on two very distinct forms, as a line to accentuate a body line, or as standalone artwork. This is the latter form—a collection of relatively short sweeps and strokes that when combined may or may not form some type of abstract image.

Although it does take a little bit of practice to use, a Beugler pinstriping tool is much easier for the beginner to create consistent results. This tool is a small canister that is filled with paint and a built-in wheel that transfers the paint to the surface. The width of the wheel determines the width

Regardless of how trim attaches to your hot rod body, it is usually a good idea to have an assistant with you when you are installing it. Since trim usually goes on after the paint is on and buffed out, you would sure hate for the other end of a long piece of trim to be scratching the paint while you are installing one end of it.

The other form of pinstriping is the red line that follows the body line on this blue roadster to add just a bit of contrasting color.

Prior to installing trim, take the time to have it replated if necessary or cleaned and polished so that it looks its absolute best when it is installed.

Make sure that all of the clips and retainers are on hand, and that you know how they work, before attaching trim pieces. As you did for their removal, have a helper assist you in replacing extra long pieces. This will not only help to prevent bends or wrinkles on the trim, but it also adds more control to the installation to prevent accidental scratches on paint finishes.

Since door handles and key locks attach directly to painted body panels, you must install them with care to avoid causing scratches, chips, or nicks to the finish. In many cases, gaskets or seals are designed for placement between hardware and body skin. If the old gasket is worn, cracked, or otherwise damaged, do not use it. Wait to install that handle or exterior door item until a new gasket is acquired.

The screws, nuts, or bolts used to secure door handles are normally accessed through openings on the interior sides of doors. You have to reach through with your hand to tighten the fasteners. Be sure to use wrenches or sockets of the correct size to make this awkward job as easy as possible. After handles and key locks are secured, you must attach linkages or cables that run to the actual latch mechanisms.

As with other exterior trim pieces and accessories, you should take this opportunity to clean, polish, and detail grille assemblies while they are off your car. Touch up paint nicks, clean tiny nooks and crannies, and wax metallic parts as necessary. Use a soft toothbrush and cotton swabs to reach into tight spaces. Should painted parts look old and worn, consider sanding and repainting them. Tiny chips or nicks can be touched up with the proper paint, using a fine artist's paintbrush.

## Washing, Polishing, and Waxing

Check with the source of your paint products to see how long you should wait to wash and wax your freshly painted hot rod. Car wash soap products can be found at auto parts stores and discount department stores. For the most part, any brand of car wash soap should be well suited for the finish on your vehicle. Just be sure to read the label for any warnings and to follow the mixing directions on labels of any product that you use.

The best way to prevent minute scratches or other blemishes on paint is to wash the vehicle in sections. Wash the dirtiest parts first, like the rocker panels, fender well lips, and lower front and rear end locations. Then, thoroughly rinse your soft cotton wash mitt and wash soap bucket. Mix up a new batch of wash soap to clean the vehicle sides. If their condition was relatively clean to start with, you can continue with that bucket of sudsy water to wash the hood, roof, and trunk areas.

This process rids your wash mitt and bucket of dirt and other scratch hazards, like sand and road grit. If you were to wash your entire car with just one bucket of sudsy water, you increase the chances of your wash mitt picking up debris from the bucket where it will then be rubbed against the vehicle's lustrous finish. Likewise, any time you notice that your wash

mitt is dirty or if it should fall to the ground, always rinse it off with clear water before dipping it into the wash bucket. This helps to keep the wash water clean and free of debris.

To clean inside tight spaces, like window molding edges and louvers, use a soft, natural hair floppy paintbrush. Do not use synthetic-bristled paintbrushes because they could cause minute scratches on paint surfaces. In addition, wrap a thick layer of heavy duct tape over the metal band on paintbrushes. This will help to guard against paint scratches or nicks as you vigorously agitate the paintbrush in tight spaces, possibly knocking the brush into painted body parts such as those around headlights and grilles.

This is a good place to discuss the difference between wax and polish, both of which you may use to maintain your hot rod's finish. Each of these products has its own separate purpose. Polishes clean paint finishes and remove accumulations of oxidation and other contaminants. Since your hot rod has just been assembled and painted, you should not need to use any polish to remove oxidation. On the other hand, wax does no cleaning. It does, however, protect those paint finishes that have already been cleaned and polished. Simply stated, polish cleans—wax protects.

Auto body paint and supply stores generally carry the largest selection of auto polishes and waxes, although many auto parts stores stock good assortments. Every polish should include a definitive label that explains what kind of paint finish it is designed for—for example, heavily oxidized, mildly oxidized, and new finish glaze. Those designed for heavy oxidation problems contain much coarser grit than those for new car finishes.

Along with descriptions of just which kind of paint finish particular polishes are designed for, labels will also note which products are intended for use with a buffer. Those with heavy concentrations of coarse grit are not recommended for machine use. Their polishing strength, combined with the power of a buffer, could cause large-scale paint burning problems.

Carnauba wax is perhaps the best product to protect automobile paint finishes. Meguiar's, Eagle 1, and other cosmetic car care product manufacturers offer auto enthusiasts an assortment of carnauba-based auto wax products. There are other paint protection products available that profess to work like wax but contain different chemical bases, which you must clearly understand before applying them to your new paint job.

Some of these (typically, they have poly or polymer in the product name) are loaded with silicone materials. Although the products may protect your car's finish for a long time, professional auto painters advise against their use because the silicone content is so high and saturating that any repainting that may be required in the future could be plagued by severe fisheye problems. In some cases, silicones have been known to penetrate paint finishes to eventually become embedded in sheetmetal panels.

# PROJECT 12
# Base Coat to Pinstripe

I have to admit, now that the bodywork portion of the track T project is complete, I am getting excited about getting it painted. No doubt, when you get to that point on one of your projects, you will, too.

**1**

With the sealer on the track T, I finally feel like I am making some progress on it. For this stage of the painting process, I have the lightweight body resting on a couple of metal sawhorses. Having the body elevated will allow me to spray the underside as the rest of the body is being sprayed. With the benefit of experience, I would recommend tilting this type of body up on the firewall to paint the underside. This does require two separate spraying operations, but I think it would have worked better.

### Disassembly

On larger, fat fendered hot rods, it may be more practical to leave the body on the chassis, mask everything off, and paint the body. On the track T, however, it is easier to remove the body than to mask the items that are not to be painted. The track T does not have opening doors, so they do not pose a problem to painting. Most painters would suggest that on a hot rod that does have opening doors, it is easier to fit and latch the doors while they are in primer and leave them on while painting the body. This may take extra effort in terms of masking, but it is less likely to require touchup. Most other body components can be removed and reinstalled after being painted. This allows for more complete paint coverage of all components. For the track T, the list of body components to be painted includes the body, left and right body aprons, deck lid, a three-piece hood, the nose, two front shock mount/headlight mount brackets, two headlights, and the rolled rear pan.

To make reassembly much easier, go about disassembly with some restraint and sensibility. Label wires and cables that will need to be disconnected. Package bolts and other fasteners that are removed in sealable bags or boxes and label them appropriately. No, you will not remember where they all go when you get ready to reinstall them if they are not labeled.

**2**

Although I had not been so anal about it during the building stage, I finally realized that disassembly prior to painting should be more organized. With a small box of resealable storage bags and a permanent marker, the various fasteners, nuts, bolts, and other small pieces are safely stored and labeled for reuse. This should make reassembly quicker and easier, as I should not need to run to the parts store to replace missing hardware.

If you are going to be removing the body from the chassis for paint, the following tip will prove to be a good one to you, especially if the body mount holes have been elongated during previous body mounting sessions. Now that the body and all of the related sheetmetal or fiberglass components fit together properly, drill two small holes through the floor into the top of the frame rail on each side of the vehicle. The farther apart these are the better, but one near the front of the (front) door and another somewhere behind the back of the back door will be sufficient. Drill these four holes so that an 8- or 16-penny nail will just slide in smoothly. During the process of reinstalling the body on the chassis, the tighter tolerance of the nails will ensure that the body is more accurately located (compared to relying just on the larger body mount holes). When all four nails slide into place easily, the body is properly located and you can bolt it down securely, knowing that it is mounted in the correct location.

## Masking

If you are going to be painting in a garage that is actually used to store vehicles, lawn mowers, and anything else that may need to be used during your painting operation, temporarily move them to a different location. Sweep or vacuum the floor to get rid of as much dust and dirt as possible. Use an air nozzle to blow away any dust and dirt from the hot rod body and related components that are soon going to be painted. Use appropriate-sized masking tape and masking paper if necessary to protect any areas that are not to be painted. From the various bodywork priming sessions and the application of sealer, you should know exactly what needs to be masked on your particular hot rod project.

Remember that the time spent masking is significantly more conducive to a job well done than the time spent cleaning up areas that were not masked properly. For the track T, the body had been removed from the chassis, so there was no masking required.

## Applying the Base Coat

As the various components are masked, situate them in your painting space (paint booth, garage, whatever) so that you can maneuver your spray gun sufficiently to cover all surfaces and also so that you will not be dragging your air hose across any parts. Most paint shops position bodies on a roll-around cart and hang all of the other parts for painting. I wish that I had built a body dolly, as that would have let me move the body to the middle of the garage to work on it and then roll it to one side of the garage at the end of the work session. That would have kept my wife's car inside at night, but luckily, she didn't complain. Welding wire works real well to hang parts that have at least a small hole in them. If you do not have anything to hang parts from, you can easily make a framework out of electrical conduit and then attach it to a couple of small roll-around feet to make it portable. Depending on how you make this, it can be easily disassembled and stored without taking up much room. Another method that is suitable for small parts is a piece of electrical conduit between two step ladders.

For my big adventure of painting the track T, I mounted the body on a pair of sawhorses, which required me to paint the underside in a couple of different settings to be able to cover everything. The sawhorses did provide me with better access to the lower edge of the body, so I did not have to bend over too much, nor stretch too much to reach the upper portions. Remember ease of reach when you are situating a car body to paint it, as you would be surprised how much a paint gun full of paint seems to weigh when you hold it at arm's length for a while. Since I didn't really have anything to hang the other parts from, I made good use of a folding work stand, a wooden stool, and some empty drywall mud buckets. While these improvisations worked well, I was concerned with how to situate the lengthy body aprons, which had different width flanges and required paint on three different surfaces.

By clamping a spring clamp onto the work stand, I was able to prop up the body aprons so that they were accessible on all sides.

I've probably said this before, but before you leave the paint supply emporium, ask for instruction sheets for each of the various primers, base coats, clears, and any other products you may use. These directions provide mixing ratios, nozzle sizes, air pressure to spray, flash time, dry times, and a wealth of other info that if followed will yield a better finish. Although the temperature outside the garage was not quite 50 degrees Fahrenheit, the well-insulated walls and having an electric heater turned on prior to painting kept the inside of the garage between 65 and 70 degrees Fahrenheit—a pretty decent temperature range for painting. For the PPG Deltron® 2000 DBC Harvest Moon (almond) base coat, directions said to mix it 1:1 with the DT reducer best suited to shop conditions, which in my case was DT870. As with most base coat paints, the directions suggest two to three coats or until hiding is achieved, allowing 5 to 10 minutes between coats. Three coats seem to have covered the black sealer sufficiently.

Looking back, I should have spread drop cloths over my roll-around tool chest, welder, grinding stand, and various other stuff in the garage, including the floor. Most of the overspray dried before reaching any surfaces, but you would be amazed at how filthy dirty it is and will continue to be until I get the garage cleaned up.

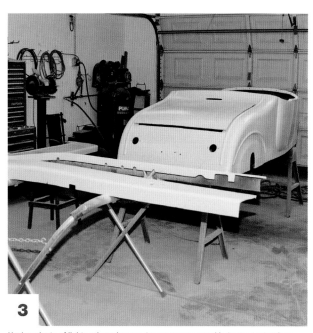

**3**

Having plenty of light and ample room to maneuver around between parts when painting is critical. I did not really have a good place to hang any of the parts for painting, so they are all situated on a variety of work stands, sawhorses, and other impromptu devices. As long as you are wearing a proper mask when spraying, you can keep the doors of the garage closed, but you should open them a bit between coats just to avoid an unhealthy buildup of fumes.

### Applying Clear

The application of clear is no different from any other paint, except that it dries to a nearly invisible finish. Because of this, you must maintain a closer eye on your work so that each pass is made uniformly. As long as you can avoid drips, sags, and runs, you are OK, as the clear will be wet sanded later to eliminate the orange peel anyway. Someone recently told me, "Lay on the clear the way you want it to look." What they meant was to avoid runs but to also avoid dry spots. The base coat itself has no shine to it once it dries. When you begin spraying clear (after the appropriate drying time), you will notice a distinct gloss to the surface. Any areas that do not have this gloss have less clear on them.

Instructions for mixing solvents and hardeners with clear paint material are provided on product labels, application guides, and information sheets. Of utmost importance is the flash time between the last color coat and application of clear. Spraying clear coats on before the solvents in color coats have sufficiently evaporated will cause checking and crazing.

I used PPG's Concept® 2021 Urethane Clear, which is mixed 4:1:1 with the appropriate temperature DT reducer and DCX hardener. While it is not necessary to have the garage door open while spraying the base coat, it was strongly suggested by those in the know to open the big door about 9 inches and have a fan blowing exhaust out the walk-through door while spraying the clear to make sure that overspray didn't end up falling onto itself. This did have me watching the garage temperature pretty closely, as paint products simply do not cure properly below 55 degrees Fahrenheit.

Four coats of clear should be plenty to allow me to wet sand the orange peel off, to provide a smooth surface for the scallops. After the scallops are applied, three to four more coats of clear will be applied and wet sanded again.

**5**

Spray the clear on the way you want it to look. Avoid runs, drips, and other errors, but other than that, you cannot really put on too much. Just make sure that you apply multiple lighter coats with the required time between them, rather than one heavy coat that will not cure properly.

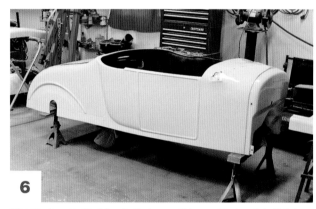

**6**

After allowing a couple of days for the base coat/clear coat to dry substantially, I moved the various components around to allow my wife to park in the garage again. This is one of the reasons a roll-around body dolly would be a good idea. Since I didn't have one, I had to ask one of the neighbor kids to help me move the body.

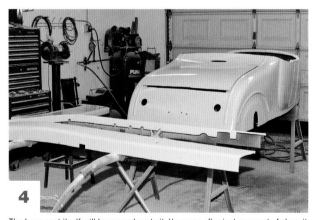

**4**

The base coat itself will have no gloss to it. However, after just one coat of clear, it is easy to see the gloss begin. If the surface is not glossy, there is no clear on it, simple as that. My local paint dealer strongly suggested that I have the garage door (or a window) open slightly with a fan blowing out the other end of the garage when spraying the clear, to keep the fumes from the clear from simply falling onto the surface.

**7**

From this angle, my first paint job does not look too bad, if I say so myself. Other than paint prep, there is no bodywork from the doors back, which is a good thing. The paint seems to cover evenly and the clear looks even, although it still needs to be wet sanded at this point.

**8**

The aluminum hood and hood sides look to be covered sufficiently. Like the rest of the components, these pieces need to be wet sanded as well, prior to reassembly and masking for scallops.

## Rework

The rules in this book do not come down from on high only for the inexperienced. Imperfect prep work can plague anyone, and when I put base coat on the track T, I discovered flaws in my bodywork in the cowl area. During the sanding and priming process, it looked OK. Yet when the base coat went on, it became evident that I had not feathered out some body-filler edges fully. I thought maybe no one else would notice, but when I put on some clear—forget it—the flaws were unacceptable. So . . . now, there is an unplanned section in this how-to book.

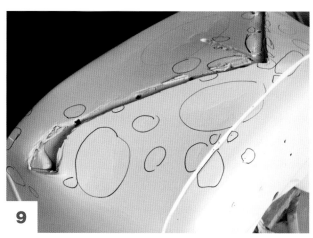

**9**

Holy smokes!!! While some of the bodywork deficiencies are not noticeable in the photos, several of them are, so you can imagine how bad they look in person. To make sure that I got them all corrected and made my point for the photo, I circled each bodywork mistake. Again . . . bodywork and paint prep will make or break a paint job. Painting over crap will simply yield painted crap.

Each paint manufacturer has its own recommendations on how to go back in, lightly fill, and then feather these edges. The processes depend mostly on the type of paint that you have used.

For my situation, there were multiple defects in the relatively small area of the cowl. I began by sanding the overall area with 180-grit sandpaper without water to scuff the surface. The next step was to mix a small amount of body filler and very carefully apply it to the areas that had not been feathered properly. After the filler dried, I sanded it smooth with 120- and then 180-grit sandpaper until any defects were gone. The next step was to mask off the body so that no stray guide coat, primer, or paint would cause additional repairs. Leaving several inches of "blending" room around the repairs, I covered the rest of the body. I then used some spray can enamel to add guide coat to the repaired areas. That was sanded off with 220-grit sandpaper and then followed by 400-grit sandpaper. I then applied two light coats of sealer to the areas of new body filler, and then added a third to the overall area.

**10**

Hanging my head in shame, I began sanding the cowl with 180-grit sandpaper to scuff the clear and base coat so that the body filler will adhere.

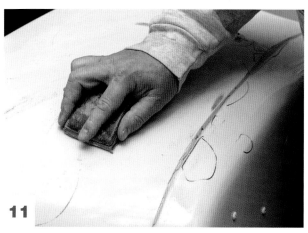

**11**

Although there were several individual errors, the small size of the cowl on the track T necessitated that the entire cowl area be sanded.

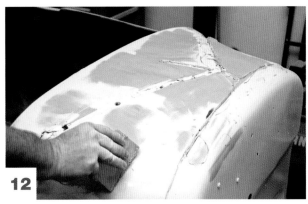

**12** Likewise, a thin coat of body filler was applied to most of the top of the cowl as well. The V-shaped area without filler is where the windshield frame fits the body.

**13** While waiting for the body filler to set up, I went ahead and masked off the rest of the body. I did not want to get any guide coat sprayed anywhere that I was not planning on doing any sanding.

**14** I began sanding the new body filler by using 120-grit sandpaper wrapped around a squeegee to serve as a sanding block. Having gotten into this rework predicament by going too fast in the first place, I made sure that I took my time and got all of the voids filled and the edges feathered out properly.

**15** A little more filler was necessary to get the cowl smoothed out completely; however, these are very thin applications at this point. Although you still have to use sufficient hardener to get the filler to set, you can mix it a little cooler, so you have a little more time to spread it out the way you want before it begins setting up. I sanded the new filler with 120-grit and then sanded the entire area with 180-, 220-, and 400-grit sandpaper.

**16** When I was satisfied with the reworked area, I used an air hose to blow off any sanding dust, then wiped the area down with wax and grease remover. Then I mixed up a small amount of sealer and applied a couple of coats. The sealer is necessary to prevent any of the filler from bleeding through to the paint.

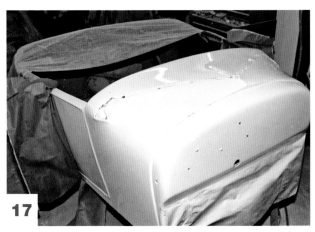

**17** After sufficient drying time for the sealer, three coats of base coat were applied, along with three coats of clear. I did peel the masking back from the doors a little bit in order to blend the color and clear into the old. The cowl looks much better this time around.

After waiting overnight to ensure that the sealer had dried sufficiently, I applied the Harvest Moon base coat with sufficient coats to achieve coverage. Each coat went slightly past the previous coat to blend into the original base coat. After the proper dry time, three coats of clear were sprayed over the entire area, again spreading slightly past the previous application, but not expanding all the way to or past the masked edge.

Although this "issue" caused some delay (both getting scallops on the track T and getting this manuscript finished), you can bet that I'll make sure that I do not make this same mistake again. It will most likely be something totally different. . . . Sadly, making mistakes is how we learn much of what we know, so think of it as a positive experience and move on.

### Color Sanding

The term "color sanding" probably gets its name from back in the days of lacquer paint. When lacquer paint was the rage, several coats of color would be applied, with each coat meticulously wet sanded before successive coats were added. This labor-intensive process yielded some flawless paint jobs, as each application of color was sanded to perfection. With contemporary paint products, most notably base coat/clear coat, the color is not sanded, but the clear is sanded to provide the best gloss.

Now that the clear had had plenty of time to cure, it was time to wet sand it to remove any orange peel. Orange peel gets its name from the fruit and is a paint condition with a very similar look. While the clear coat is what provides the gloss to the otherwise dull base coat, the surface should be as smooth (and flat) as possible to have the best gloss. Flat finishes shine the most because they reflect light so directly. A mottled surface, like an orange's peel, deflects light in many directions and therefore looks duller. By sanding away the clear's orange peel texture, the painted surface can become optically flat and provide excellent gloss.

Use a rubber squeegee as a sanding block, plenty of water (with a bit of car wash soap to act as a lubricant), and extremely fine sandpaper (1,500- to 2,500-grit wet-or-dry) to remove imperfections in the surface of the clear. Only slight pressure is required and the sandpaper should be moved in a circular pattern. Be sure to keep the surface wet and do not sand through the clear and into the color below. If you do sand into the clear, it's easy to add additional coats of clear, but sanding into the color becomes a repair operation that is best avoided.

### Reassembly

By now, there is some light at the end of the hot rod building tunnel. All of the parts that were bare steel and raw fiberglass have now been fitted up, sanded down, prepped, and painted in the desired color. The masking material is removed and reassembly can begin. For you, much of this will be installing

**18** This rubber squeegee is what I used as a sanding block during the "rework" of the cowl and the wet sanding of the clear. The opposite side of the squeegee does not have any embossed lettering on it, so that side is used against the surface being sanded. When wet sanding, a circular motion should be used, with plenty of water.

the various pieces and parts that are not sheet metal, such as lights and trim. If flames, scallops, or any artwork is to be added, assembling the sheetmetal components is simply the next step toward an additional paint job. To prevent misaligned graphics or artwork, all panels must be installed and properly aligned. Hinges and latches should be installed (if not already) and adjusted as required.

For a closed vehicle, or if graphics are not going to be applied immediately, reassembly can be completed to the point of making your new hot rod drivable. For an open hot rod such as my track T, most of the body components (hood, nose, body aprons, and body) will be removed for painting for ease of both accessibility and masking. For this reason, only the body components will be reassembled prior to application of the scallops. Wiring, steering, and the other assembly that must be done prior to driving will be completed later.

### Laying Out the Scallops

As seen earlier in this book and discussed in my book *How to Build a Cheap Hot Rod*, Steve Gilmore of Stilmore Designs provided several layout designs for the scallops. With a photocopy of the chosen layout and a couple of rolls of 3M ⅛-inch Fine Line tape in hand, I was ready to begin laying out the scallops. A few other things that came in handy during the layout process were a water-soluble marker, a tape measure, and a flexible ruler. Make sure that any marker that you use for making reference marks is water soluble so that those marks can be removed from the painted surface easily. There are probably other ways to lay out scallops and I certainly do not do this professionally, but the process worked for me.

Begin by positioning the vehicle on a level surface and in a place where you can back away from it to see how your layout really looks. If the vehicle is not parked on a reasonably level surface, it will be difficult to layout the scallops correctly as your brain will be trying to compensate for level. You need to be able to step back from the layout so that you can gain a better perspective, rather than being forced to have a skewed view from being too close.

**19**

Although you will still need to outline flames or scallops with Fine Line tape, being able to draw a design or make reference marks on the painted surface will make the task easier. A water-soluble pen (commonly referred to as a sign pen) is available in many colors and can be wiped off the painted surface with a damp cloth (or some spit and your thumb). Just be sure to test the ability to remove the pen's ink in an inconspicuous location on the car first.

**20**

Before any artwork (flames, scallops, or any other artwork that spans multiple body panels) can be laid out, the body components should be assembled. This includes being hinged and latched, or otherwise secured in place when it comes to doors, hoods, and deck lids.

Establishing the baseline was the most difficult part, as there are few if any parallel lines or square surfaces on the body and hood. After a few tries, I was able to lay down a line of ⅛-inch Fine Line tape that is roughly parallel to the top of the hood side. By placing one end of the tape on the body and holding it with one hand, and then pulling the tape roll with the other hand, I was able to keep the tape straight. Be careful to not stretch the tape, but keep it taut. Press the tape onto the body and especially into irregular shapes, such as around the imitation door outline so that the tape has full contact with the body. Otherwise, paint will find its way between the Fine Line tape and the body.

**21**

Establishing a baseline for the scallops was the most difficult part of the job. Flame layout usually begins in the middle of the hood and works outward, but with the geometric pattern of scallops, it made more sense to start with the sides first.

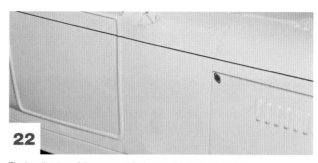

**22**

The baseline (top of the upper scallop) seemed to look the most appropriate being roughly centered between the top of the hood side panel and the top edge of the door. With the rounder shapes of older car bodies, finding similar locations to measure from to determine a set of parallel lines can be difficult. This makes stepping back to take a look at the overall picture even more critical.

With the baseline (the top edge of the top scallop) looking to be in the correct location, I measured down some distance for the top of the middle scallop and that same distance again for the top of the bottom scallop. Unlike flames, scallops are a geometric design. That does not mean that distances must be equal, but if not, they are usually proportional for the best results. With two sets of points marked, I used Fine Line tape to connect the dots, creating the top line of each scallop.

**23**

Now that a baseline has been established, there is something to measure from to establish parallel lines. Two more evenly spaced points were marked on the cowl and near the back of the passenger compartment. These marks would establish the top line of the middle and lower scallop. A flexible steel ruler with a cork back is being used in the photo, but a cloth tape like a seamstress would use will be less likely to scratch any paint.

**24**

With the small size of the track T, stretching tape is not a big deal, but on a larger vehicle, assistance might be necessary. Even if you can reach both ends of the tape, a second set of eyes can be useful in making sure everything looks good.

The next step was to establish a baseline for the bellies of the scallops. This line can be straight, a circular curve, or an irregular curve, but it should be something that can be duplicated on the other side of the vehicle. Using my Steve Gilmore artwork as a guide, I chose two points on the profile of the car and used a piece of Fine Line tape to create a baseline between the two. Admittedly, the location of this baseline is quite arbitrary, but like many things, you have to make a decision and move on.

**25**

The diagonal tape is the baseline for the bellies of the scallops. Its location is arbitrary, but it should be located somewhere that is easy to re-create on the opposite side of the vehicle.

The next steps will determine the lines that represent the bottom of each scallop. Many painters have fallen down on this by not getting the lower edges parallel to each other, even though the top lines are. Begin by measuring along each top line from its intersection with the belly baseline the same distance toward the back of the vehicle and marking

a point with a water-soluble pen. Using a pen to mark reference points is much easier than using strips of tape that get in the way when you actually start masking. Now to get the angle the same on each scallop, measure downward along the belly baseline from its intersection with the top line of each scallop some distance. Now use Fine Line tape to connect the dot at the baseline with the dot representing the point of the scallop. Since the baseline is at a pretty steep angle with the top lines, this distance will be quite a bit more than the final width of the scallop. As I recall, I measured 5 inches along the baseline, which ended up with the scallops being about half that width at their widest point. Again, this distance is arbitrary, so you will need to lay it out, step back and look, and then adjust accordingly.

**26**

By measuring from the intersection of the top line of each scallop with the belly baseline, the angle of the lower line of each scallop can be determined. For the angle to be identical, each measurement along the upper line must be the same and each measurement along the belly baseline must be the same.

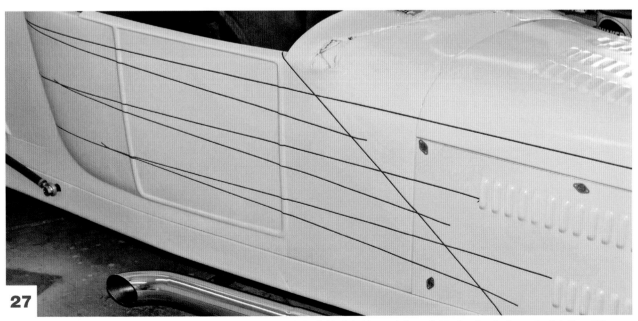

**27**

It should be obvious that some of the tape will be trimmed off prior to the layout being finished. Just be sure to leave all of the baselines in place on the first side until the layout is re-created on the opposite side.

Bellies of scallops can come to a single point, be angular, circular, or curved. To match my aforementioned artwork, I simply used a French curve to define a curve from the lower edge of the scallop to a point being the intersection of the belly baseline and the top line of the next scallop. To make sure that the curve is similar on each scallop, I made reference marks on the French curve so that it was positioned the same on each one. You could certainly use a pattern made from poster board to serve the same purpose. I know of at least one set of scallops on a '32 Ford that used a portion of artwork from a case of Bud Light as a pattern.

With the side of the scallops laid out, it was time to move to the hood top. The intent is to have the edge of one scallop running up the center of the hood and then curve around to blend in with the top line of the side scallop. Another goal (criterion, actually) is to not have any louvers passing through the edge of the paint. Additionally, I did not want any painted edges crossing the front portions of the hood and nose area, due to fear that vibration-induced movement in that area would result in misalignment of the artwork. The back of the hood is secured to the body, so movement is minimal there, but the front end of the hood and nose are not as static. After multiple attempts, I was able to freehand a line that blended into the side adequately and met the requirements that I had imposed.

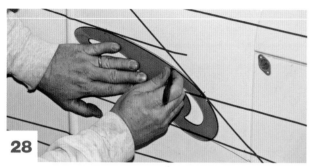

**28**

Although it would be easy enough to lay tape out along an arc at the belly of the scallop freehand, I did not feel confident in being able to repeat the same shape adequately enough. By tracing around a French curve with a sign pen, I could establish a guideline for the tape to follow. Reference marks made on the French curve were used to align it correctly with each scallop.

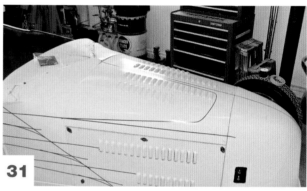

**31**

The first attempt at laying out the hood top scallops was not too bad on the top but created too much of an angle where the hood and the cowl come together. I could have started the transition closer to the front of the door and began moving away from being parallel with the hood side as the tape moved forward, but I wanted the line to be more parallel with the hood side and louvers.

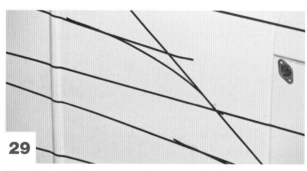

**29**

After tracing around the French curve with a sign pen, Fine Line tape was applied over the line.

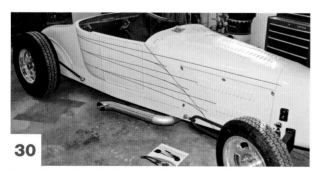

**30**

This is what the side scallop layout looked like before removing any of the extra tape. Having a drawing to use as a design aid is a great help when laying out scallops.

**32**

Pulling the line closer to the side of the hood in front made the transition much smoother with no discernable angle in the middle. Sharp angles on a round-bodied car usually do not look right.

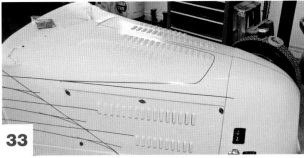

**33**

One side down, another side to go. I re-created the pattern on the opposite side by using guidelines and measurements. Another method is to cover the artwork with Kraft paper, tape it down, make some reference marks, and then trace over the Fine Line tape with a pounce wheel. The pounce wheel will make a series of small holes in the Kraft paper. After repositioning the Kraft paper on the other side of the vehicle, carpenter's chalk line chalk can be spread over the paper. When the paper is removed, the chalk will be along the line to be masked.

After stepping back and looking at the layout for a while, I declared it good and then proceeded to duplicate the layout on the opposite side. Make sure that you do not remove any of the extra Fine Line tape used for baselines until you have measurements duplicated on the opposite side. To re-create the design on the hood top, I used a piece of drafting vellum. I taped it down over that portion of the hood top, marked several reference points on the vellum that could be matched with points on the other side of the hood, and then traced over the Fine Line tape. I then removed the tape and cut along the line just made on the vellum. Make sure that your reference marks will still be attached to the vellum when you cut along the line. Now position the vellum pattern and tape it in place. Using a water-soluble pen, trace along this pattern onto the hood. Now the pattern can be removed and Fine Line tape used to outline the scallop. When the layout is complete, remove any extraneous tape that is not part of the final outline. Press down all of the tape firmly so that it does its job of masking, but make sure that you do not move the tape out of position in the process.

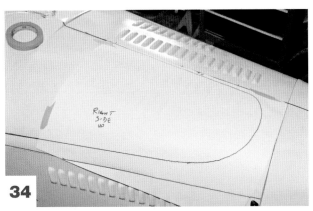

**34**

I had no feasible way to lay out this curve geometrically, so after doing it by hand on one side, I used vellum tracing paper to create a pattern by tracing over the Fine Line tape, cutting out the pattern, and then tracing around it.

**35**

Removing extraneous Fine Line tape allowed the first real look at the scallops. Perhaps not perfect, but not too awful. Good, bad, or otherwise, I can say I did it myself. If I can, most likely you can, too.

### Masking the Scallops

I used ¾-inch-wide automotive-grade masking tape to do most of the masking. Begin by placing one edge of the masking tape on the Fine Line tape, with the width of the masking tape extending toward the area that is to be masked. The Fine Line tape is more flexible and will actually define the edge of the painted scallops. The masking tape should not extend past the edge of the Fine Line tape; however, you must avoid leaving slivers of unmasked area between these first two pieces (or any other pieces) of tape. Go around the entire outline of the Fine Line with the masking tape. Make sure that you press the tape down firmly. I used masking tape between the scallops and then used a patchwork of masking paper to cover the larger areas.

**36**

Fine Line tape is more flexible than masking tape, so it is used to lay out the scallops and provide the actual masked edge of the painted surface. Whether using more masking tape or using masking paper, the Fine Line tape should be followed by one line of ¾-inch masking tape. This is flexible enough to stick to the ⅛-inch Fine Line tape but wide enough to make attaching more tape or paper possible.

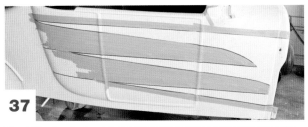

**37**

The additional application of masking tape also provides a more visual look at the layout. Now is the time to change the layout if anything is not the way you want it.

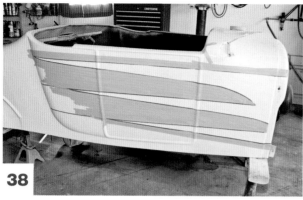

**38** Due to the irregular shape, I chose to use masking tape to mask the area between the scallops. The one roll of 18-inch-wide masking paper that I have would have required too much cutting and trimming in this area. I did use masking paper on the back half of the car and along the lower edge, however.

### Scuffing the Clear

Since the scallops will need to be scuffed prior to applying paint, another strip of Fine Line tape should be applied just inside of the original line of Fine Line. This will allow you to use a red Scotch-Brite pad to aggressively scuff the bulk of the scallop area, while providing protection for the Fine Line tape that represents the edge of the layout. When the bulk of the area is scuffed, remove the second line of tape, and carefully scuff the edges of the area that is to be painted.

Whether you mask and then scuff, or scuff and then mask is a matter of personal preference. Just make sure that you scuff the surface to be painted adequately so that the paint sticks. Make sure that you mask fully so that you do not have slivers of paint in the wrong places. However, if slivers of paint appear when everything is said and done, they can be removed with some polishing compound if there is clear between them and the underlying paint.

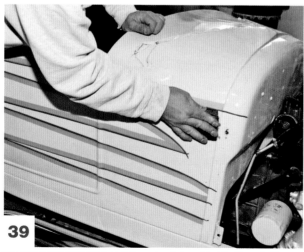

**39** Use a red Scotch-Brite pad to scuff the surface to be painted. Scuffing will allow the paint to adhere better. You should also clean the surface with wax and grease remover prior to painting it.

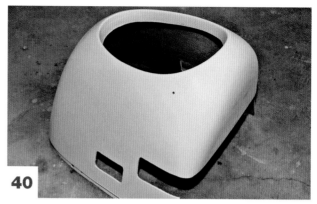

**40** The entire nose, to be painted the color of the scallops, has been scuffed. You may not be able to tell in the photo because of the light tan color, but there is very little if any reflection in the paint after being scuffed. Have no fear, it is much easier to see in person when you are working on your hot rod.

**41** It may be a little easier to see the difference between the scuffed and the nonscuffed areas of the hood top. The area on the front, center, and edges has been scuffed and will be painted. The area that will not be painted has had the clear wet sanded, so it is a bit glossier. When all of the painting is done and the entire hot rod buffed and polished, it will all be very shiny.

### Painting the Scallops

After double-checking that everything is masked sufficiently and that you have adequate working space around everything that is to be painted, the painting of scallops can begin. As with all base coat paint products, color is applied as required to provide complete coverage. The scallops for my track T are Medium Mocha, which is a dark brown that matches the single stage color used for the chassis and engine, albeit in a base coat formula. The base coat/clear coat paint system used for the body and scallops proved to be more user friendly than the single-stage paint followed by clear used on the chassis. Whether using a single stage or multistage paint system, you should always apply multiple coats of clear prior to adding artwork.

### Applying Overall Clear

After waiting the recommended dry time, clear was applied with the proper flash time between coats. As the clear begins to dry, it will immediately begin to turn glossy. If the surface is not glossy after the second coat of clear is applied, you are not getting coverage. Although the clear has no pigment, it does begin to "gloss" fairly quickly, which tells you if you missed any spots. Apply three or four coats of clear to make sure that you have enough material to work with in the buffing process.

Now that the scallops are on the track T, the panels that received scallops will be wet sanded to remove any orange peel. And then, the buffing and polishing process will begin, followed by reassembly and many miles of open air hot rodding. I hope to see you on the road . . .

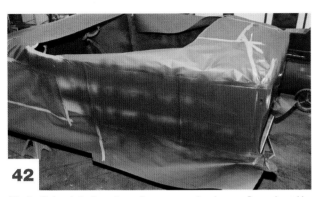

**42**

After the first coat of color on the scallops, coverage is only so-so. Remember, add enough coats of base coat to achieve complete coverage, allowing sufficient flash time between coats. The base coat itself will look dull.

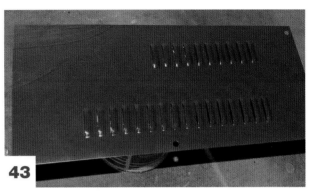

**43**

All but a small portion of the hood sides will be painted the scallop color. If you look close, you can still see the masked off area after this first coat of color.

**44**

Although it looks very white in the photos, the body is VW Harvest Moon, which is a light tan. Buffed, with an orange pinstripe added between the two colors, and sitting in outdoor light, the track T should look very traditional. Notice that the squared-off belly between the middle and lower scallop will be hidden when the hood side is in place.

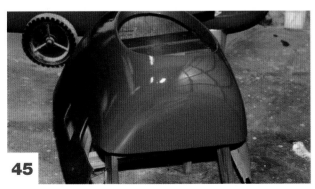

**45**

After applications of both base coat and clear coat, the nose looks pretty good as it is. In all actuality, all of the body components will be wet sanded with 1,000-, 1,500-, and 2,000-grit sandpaper, and then buffed and polished to bring out the best shine possible.

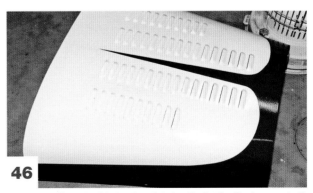

**46**

After being buffed and polished, a ⅛- or ³⁄₁₆-inch-wide pinstripe (Omaha orange) will finish off the scallops.

# Sources

**Air Ride Technologies**
350 S. St. Charles Street, Jasper, IN 47546
www.ridetech.com
812-481-4787
*Air suspension components*

**Campbell Hausfeld**
100 Production Drive, Harrison, OH 45030
www.chpower.com
888-247-6937, 513-367-4811
*Air compressors and pneumatic tools*

**Car Quest of High Ridge**
3032 High Ridge Boulevard, High Ridge, MO 63049
636-677-3811
*Auto body repair supplies*

**Chief Automotive Systems**
1924 E. 4th St., Grand Island, NE 68801
www.chiefautomotive.com
877-644-1044
*Chassis and unibody straightening equipment*

**Danny Miller's Rear Gears**
921 St. Louis Ave., Valley Park, MO 63088-1933
www.rear-gears.com
636-861-3900
*Rear axle housings, gears, brakes*

**Dagger Tools**
47757 West Rd., Unit C-106, Wixon, MI 48393
www.daggertools.com
248-735-1123
*Metal-shaping tools*

**Eastwood Company**
263 Shoemaker Road, Pottstown, PA 19464
www.eastwoodcompany.com
800-345-1178
*Automotive restoration tools, equipment, and supplies*

**Hemmings Motor News**
P.O. Box 100, Bennington, VT 05201
www.hemmings.com
800-227-4373
*Classified ads for vehicles, products, and services*

**Howron Metal Benders**
P.O. Box 171, Aubrey, TX 76227
www.metalbendingtool.com
877-300-0171
*Metal bending tools*

**HTP America**
3200 Nordic Road, Arlington Heights, IL 60005-4729
www.usaweld.com
800-872-9353, 847-357-0700
*Welders, plasma cutters, tools, and accessories*

**Jerry's Auto Body, Inc.**
1399 Church St., Union, MO 63084
636-583-4757
*Auto body repair*

**Karg's Hot Rod Service**
6505 Walnut Valley Lane, High Ridge, MO 63049
www.kargshotrodservice.com
314-809-5840
*Hot rod fabrication and assembly, welding, painting*

**KBS Fabricators**
108 St. Joseph Ave., O'Fallon, MO 63366
www.kbsfabricators.com
636-272-1008
*Metal fabrication and louvers*

**Mayhem Custom Paint and Airbrush**
430 MacArthur Avenue, Washington, MO 63090
www.mayhemairbrush.com
636-390-8811
*Custom artwork*

**Meguiars**

17991 Mitchell South, Irvine, CA 92614-6015

www.meguiars.com

800-854-8073, 949-752-8000

*Car care products*

**Miller Electric Mfg. Company**

1635 W. Spencer St., P.O. Box 1079, Appleton, WI
54912-1079

www.MillerWelds.com

800-4-A-MILLER

*Welding power sources, plasma cutters, and welding accessories*

**Mittler Brothers Machine & Tool**

10 Cooperative Way, Wright City, MO 63390

www.mittlerbros.com

800-467-2464

*Fabrication tools and equipment*

**Morfab Customs**

79 Hi-Line Drive, Union, MO 63084

www.morfabcustoms.com

636-584-8383

*Hot rod fabrication and assembly, welding, painting*

**Mothers Polish Company**

5456 Industrial Drive, Huntington Beach, CA
92649-1519

www.mothers.com

800-221-8257, 714-891-3364

*Polishes, waxes, and cleaners*

**Pete & Jake's Hot Rod Parts**

401 Legend Lane, Peculiar, MO 64078

www.peteandjakes.com

800-334-7240

*Hot rod chassis, frames, and parts*

**PPG Refinish Group**

19699 Progress Drive, Strongsville, OH 44149

www.ppgrefinish.com

440-572-6880 fax

*Paint products*

**Ravon Street Rods**

23 High K, Unit D, Belgrade, MT 59714

www.ravonstreetrods.com

406-388-7867

*Fiberglass bodies*

**Show Me Rod & Custom**

165 Hopkins Road, Highlandville, MO 65669

www.showmerodandcustom.com

417-443-7002

*Fiberglass bodies*

**Stilmore Designs**

11458 Lucerne, Redford, MI 48239

www.stilmoredesigns.com

*Concept drawings*

**Super Bell Axle Company**

401B Legend Lane, Peculiar, MO 64078

816-758-5044

*Axles, spindles, perches, and brakes*

**The Paint Store**

2800 High Ridge Blvd., High Ridge, MO 63049

636-677-1566

*PPG paint products and supplies*

**Woodward Fab**

1480 Old U.S. Highway 23, Hartland, MI 48353

www.woodward-fab.com

800-391-5419

*Fabrication tools and equipment*

# Index

# The Best Tools for the Job.

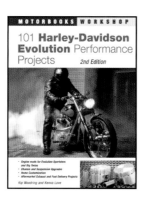

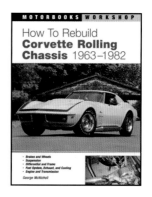

## Other Great Books in This Series